Problem Solvers

Edited by L. Marder
Senior Lecturer in Mathematics, Universit

No. 7

Vector Fields

Problem Solvers

1. ORDINARY DIFFERENTIAL EQUATIONS—J. Heading
2. CALCULUS OF SEVERAL VARIABLES—L. Marder
3. VECTOR ALGEBRA—L. Marder
4. ANALYTICAL MECHANICS—D. F. Lawden
5. CALCULUS OF ONE VARIABLE—K. Hirst
6. COMPLEX NUMBERS—J. Williams
7. VECTOR FIELDS—L. Marder
8. MATRICES AND VECTOR SPACES—F. Brickell

Vector Fields

L. MARDER

Senior Lecturer in Mathematics
University of Southampton

LONDON · GEORGE ALLEN & UNWIN LTD

RUSKIN HOUSE   MUSEUM STREET

First published in 1972

ISBN 0 04 512014 5 *hardback*
0 04 512015 3 *paper*

Printed in Great Britain
in 10 on 12 pt 'Monophoto' Times Mathematics Series 569
by Page Bros (Norwich) Ltd., Norwich

Contents

Chapter 1

Scalar and Vector Fields

1.1 Definitions A *scalar* is a quantity, like the mass of a particle, which is prescribed by a single real number; the real numbers themselves are scalars. If a scalar $\phi(x, y, z)$ takes a definite value at each point (x, y, z) in a region of space, a *scalar field* is defined in the region. Examples are (i) the pressure in the atmosphere, (ii) the density within the earth, (iii) the gravitational potential in space. It is implicitly assumed that the value of ϕ at any point is independent of the choice of rectangular coordinate system used to label the points.

A *level surface* of a scalar field $\phi(x, y, z)$ is any surface ϕ = constant. Through each point (x_0, y_0, z_0) in the region there passes a level surface, namely the surface $\phi(x, y, z) = \phi(x_0, y_0, z_0)$, and if ϕ is a single-valued function of x, y, z there is only one surface through each point.

A *vector* is a quantity, such as the velocity of a particle, which is prescribed by a positive real number (the *magnitude*) and a direction. It is denoted by bold type (**F**, **a**, etc.), and may be represented by a directed line-segment **AB**, or by three real numbers such as its components relative to a chosen system of rectangular axes. These components depend on the orientation of axes in a way described in Chapter 6. (It is assumed that the reader is familiar with the ordinary algebra of vectors.) When a vector $\mathbf{F}(x, y, z)$ is defined at each point (x, y, z) in a region, a *vector field* exists in the region. Examples of vector fields are (i) the velocity at any point in a moving fluid, (ii) the gravitational force on a particle in space, (iii) the earth's magnetic field. If **i**, **j**, **k** denote the unit vectors in the directions of the x, y, z axes, respectively, of a rectangular coordinate system, we can express the vector field in component form:

$$\mathbf{F}(x, y, z) = F_x(x, y, z)\mathbf{i} + F_y(x, y, z)\mathbf{j} + F_z(x, y, z)\mathbf{k}.$$

A curve whose tangent vector at each point is in the direction of **F** at that point is called a *field line*. Magnetic lines of force, and streamlines in a moving fluid are examples of field lines.

1.2 Scalar Fields and Level Surfaces

Problem 1.1 Give the regions in which the following scalar fields are defined: (i) $\phi(x, y, z) = 2x^2 + y^2 + 3z^2$, (ii) $\phi(x, y, z) = \cos^{-1}(x^2 + y^2 - z)$, (iii) $\phi(x, y, z) = \ln(x^2 - y^2)$.

Solution. (i) Since $2x^2+y^2+3z^2$ is real for all real values of x, y and z, the equation defines a scalar field throughout all space.

(ii) For real values of ϕ we have $-1 \leqslant \cos\phi \leqslant 1$, and so the equation defines a scalar field in the region $-1 \leqslant x^2+y^2-z \leqslant 1$, i.e.

$$z-1 \leqslant x^2+y^2 \leqslant z+1, \tag{1.1}$$

which is bounded by the paraboloids of revolution $x^2+y^2 = z \pm 1$. The field is not single-valued unless the range of ϕ is suitably restricted. For example, we may take the principal value of the inverse cosine, so that at each point in the region (1.1) a unique value of ϕ in the range $0 \leqslant \phi \leqslant \pi$ is defined.

(iii) For ϕ to be real we require $x^2-y^2 > 0$, i.e. $|x| > |y|$. This region consists of two parts, given separately by $x > |y|$ and $x < -|y|$, both of which lie between the perpendicular planes $y = \pm x$ (through the z-axis). The first part is identified as that containing the positive x-axis, and the second as that containing the negative x-axis. □

Problem 1.2 Describe the level surfaces of ϕ in cases (i), (ii), (iii) in Problem 1.1.

Solution. (i) The level surfaces are

$$2x^2+y^2+3z^2 = C, \tag{1.2}$$

where C is a constant. If $C < 0$, the locus has no real points, and if $C = 0$ it consists of a single point, the origin. If $C > 0$, (1.2) represents an ellipsoid with centre at the origin and principal axes along the coordinate axes. (The nature of the surface is most easily seen by considering the curves of intersection of (1.2) with each family of planes $x = a$, or $y = b$ or $z = c$, every real curve of intersection being an ellipse with centre on one coordinate axis and principal axes parallel to the other coordinate axes.) Putting pairs of the variables x, y, z equal to zero, in turn, shows that the major axis lies along Oy and is of length $2\sqrt{C}$, the minor axis is along Oz and is of length $2\sqrt{(\frac{1}{3}C)}$, and the intermediate axis is along Ox and is of length $2\sqrt{(\frac{1}{2}C)}$.

(ii) Equating ϕ to a constant shows that the level surfaces are of the form

$$z = C+x^2+y^2, \tag{1.3}$$

where C is a constant and $-1 \leqslant C \leqslant 1$ (by (1.1)). Since x^2+y^2 is the square of the distance from Oz, each level surface is a surface of revolution obtained by rotating the parabola $z = C+x^2$, $y = 0$ about Oz.

(iii) The level surfaces are

$$x^2-y^2 = C, \qquad C > 0, \tag{1.4}$$

and each is a hyperbolic cylinder with z-axis, formed by all straight lines parallel to Oz which meet the hyperbola $x^2-y^2 = C, z = 0$. □

Problem 1.3 (*Gradient*) Show that at each point P the vector

$$\text{grad}\,\phi \equiv \frac{\partial\phi}{\partial x}\mathbf{i}+\frac{\partial\phi}{\partial y}\mathbf{j}+\frac{\partial\phi}{\partial z}\mathbf{k}$$

is normal to the level surface of ϕ that passes through P.

Solution. Let P be the point (x_0, y_0, z_0), and let

$$\phi(x, y, z) = C \tag{1.5}$$

be the level surface through P. Consider any point $Q(x_0+dx, y_0+dy, z_0+dz)$, near P, on the tangent plane to (1.5) at P. By taking differentials in (1.5), using the chain rule,

$$d\phi = \frac{\partial\phi}{\partial x}dx+\frac{\partial\phi}{\partial y}dy+\frac{\partial\phi}{\partial z}dz = 0. \tag{1.6}$$

This shows that the vector grad ϕ (the *gradient* of ϕ) is perpendicular to $\mathbf{PQ}(= dx\,\mathbf{i}+dy\,\mathbf{j}+dz\,\mathbf{k})$, and since Q is arbitrary on the tangent plane to (1.5) at P, it follows that grad ϕ is normal to the level surface. □

Note that if $\mathbf{r} = x\mathbf{i}+y\mathbf{j}+z\mathbf{k}$, then the first of (1.6) may be written in the useful form

$$d\phi = \text{grad}\,\phi\,.\,d\mathbf{r}. \tag{1.7}$$

Problem 1.4 Find the unit normal vector, at $P(3, -2, 0)$, to the level surface of $\phi(x, y, z) = x^2+2y^2-x\cos yz$ through P, pointing in the direction in which ϕ is increasing. Calculate the rate of change of ϕ with respect to distance in this direction at P.

Solution. By Problem 1.3, a normal vector to the level surface is

$$\begin{aligned}\mathbf{n} = \text{grad}\,\phi &= (\partial\phi/\partial x)\mathbf{i}+(\partial\phi/\partial y)\mathbf{j}+(\partial\phi/\partial z)\mathbf{k}\\ &= (2x-\cos yz)\mathbf{i}+(2z+xz\sin yz)\mathbf{j}+(2y+xy\sin yz)\mathbf{k}\\ &= 5\mathbf{i}-4\mathbf{k},\end{aligned} \tag{1.8}$$

at $(3, -2, 0)$. The corresponding unit vector is

$$\hat{\mathbf{n}} = \mathbf{n}/|\mathbf{n}| = [5^2+(-4)^2]^{-\frac{1}{2}}\mathbf{n} = (5\mathbf{i}-4\mathbf{k})/\sqrt{41}. \tag{1.9}$$

Since $\pm\hat{\mathbf{n}}$ are both unit normal vectors we must determine which corresponds to the direction in which ϕ is increasing. Let $d\mathbf{r} = \hat{\mathbf{n}}\,ds$ denote a displacement of magnitude ds in the direction of $\hat{\mathbf{n}}$. Since, by (1.8), $d\mathbf{r}$ is parallel to grad ϕ (in the same sense), we get from (1.7)

$$d\phi = \text{grad}\,\phi\,.\,d\mathbf{r} = |\text{grad}\,\phi||d\mathbf{r}| = |\text{grad}\,\phi|ds, \tag{1.10}$$

which shows that the displacement gives rise to a positive increment in ϕ.

Hence the appropriate vector is $+\hat{\mathbf{n}}$, and is given by (1.9).

The required rate of change is, by (1.8), (1.10),

$$d\phi/ds = |\text{grad}\,\phi| = \sqrt{41}, \quad \text{at } P. \qquad \square$$

Problem 1.5 (*Directional derivative*) The atmospheric pressure at the point $P(x, y, z)$ in a certain region is $p(x, y, z)$. (i) Find the rate of change of pressure with respect to distance, at P, in the direction of a given vector $\mathbf{a}$. (ii) For what direction of $\mathbf{a}$ is this rate of change greatest? (iii) Find the rate of change of pressure, with respect to time, encountered by a particle which passes P with speed V in the direction $\mathbf{a}$.

Solution. (i) If $\mathbf{r} = x\mathbf{i}+y\mathbf{j}+z\mathbf{k}$, then a displacement of magnitude ds in the direction of $\mathbf{a}$ from P is given by $d\mathbf{r} = \hat{\mathbf{a}}ds$. (Note that the *unit* vector $\hat{\mathbf{a}}$ must be taken in order that magnitudes agree on each side of this formula.) Substituting in (1.7) (with $\phi = p$) and dividing by ds we get

$$dp/ds = \text{grad}\,p\,.\,\hat{\mathbf{a}}, \tag{1.11}$$

which is the formula required. The right-hand side is known as the *directional derivative* of p in the direction of $\mathbf{a}$.

(ii) Since (1.11) is the component of grad p in the direction of $\hat{\mathbf{a}}$, the greatest rate of increase of p is obtained when $\hat{\mathbf{a}}$ is parallel to grad p. (Likewise, the greatest rate of *decrease* occurs in the opposite direction.)

(iii) Putting $V = ds/dt$, we get by (1.11)

$$\frac{dp}{dt} = \frac{ds}{dt}\frac{dp}{ds} = V\,\text{grad}\,p\,.\,\hat{\mathbf{a}} = \mathbf{V}\,.\,\text{grad}\,p,$$

where $\mathbf{V} = V\hat{\mathbf{a}}$ is the velocity of the particle. $\square$

Problem 1.6 The temperature at the point (x, y, z) at time t is $\phi(x, y, z, t) = xy^2+2yzt+\sin xt$. Find the rate of change of temperature, with respect to time, encountered by a particle passing the point $(2, 3, 1)$ with velocity $\mathbf{V} = \mathbf{i}+\mathbf{j}-2\mathbf{k}$ at time $t = 0$.

Solution. Along the path of the particle the space variables may be regarded as functions of the time t, so that ϕ is a composite function of t only. By differentiation,

$$\frac{d\phi}{dt} = \frac{\partial\phi}{\partial t}+\frac{\partial\phi}{\partial x}\frac{dx}{dt}+\frac{\partial\phi}{\partial y}\frac{dy}{dt}+\frac{\partial\phi}{\partial z}\frac{dz}{dt} = \frac{\partial\phi}{\partial t}+\mathbf{V}\,.\,\text{grad}\,\phi, \tag{1.12}$$

since $\mathbf{V} = (dx/dt)\mathbf{i}+(dy/dt)\mathbf{j}+(dz/dt)\mathbf{k}$. (The partial derivatives are each formed with the remaining three of the variables x, y, z, t kept constant.) Carrying out the differentiation we get

$$\partial\phi/\partial t = 2yz + x\cos xt = 8,$$

$$\text{grad}\,\phi = (y^2 + t\cos xt)\mathbf{i} + 2(xy + zt)\mathbf{j} + 2yt\mathbf{k} = 9\mathbf{i} + 12\mathbf{j},$$

when $(x, y, z) = (2, 3, 1)$ and $t = 0$. Hence by (1.12),

$$d\phi/dt = 8 + (\mathbf{i} + \mathbf{j} - 2\mathbf{k}) \,.\, (9\mathbf{i} + 12\mathbf{j}) = 29,$$

is the rate of change of temperature encountered by the particle. □

Problem 1.7 Show that the level surfaces of the scalars $\phi(x, y, z) = y^2 + z^2 - x$ and $\psi(x, y, z) = \ln(y^2 + z^2) + 4x$ form orthogonal families.

Solution. We need to show that the normals to any pair of level surfaces, one from each of the families ϕ = constant, ψ = constant are perpendicular at each point of intersection. The normals to the respective families are in the directions

$$\text{grad}\,\phi = -\mathbf{i} + 2y\mathbf{j} + 2z\mathbf{k}, \qquad \text{grad}\,\psi = 4\mathbf{i} + 2(y^2 + z^2)^{-1}(y\mathbf{j} + z\mathbf{k}),$$

Hence $\quad \text{grad}\,\phi \,.\, \text{grad}\,\psi = -4 + 4(y^2 + z^2)^{-1}(y^2 + z^2) = 0,$

which shows the result. □

Note that, more generally, if $u(x, y, z) = a$ and $v(x, y, z) = b$ are any two surfaces which intersect, the angle θ between their unit normals $\hat{\mathbf{n}}_1$ and $\hat{\mathbf{n}}_2$ at a point of intersection is given by

$$\cos\theta = \hat{\mathbf{n}}_1 \,.\, \hat{\mathbf{n}}_2 = (\text{grad}\,u \,.\, \text{grad}\,v)/|\text{grad}\,u||\text{grad}\,v|,$$

assuming that the denominator does not vanish.

1.3 Vector Fields and Field Lines We recall that a field line of a vector field $\mathbf{F}(x, y, z)$ is any curve whose tangent, at each point, is parallel to $\mathbf{F}$. If $\mathbf{r} = x\mathbf{i} + y\mathbf{j} + z\mathbf{k}$ is the position vector of any point on a field line, then the differential displacement vector $d\mathbf{r} = dx\,\mathbf{i} + dy\,\mathbf{j} + dz\,\mathbf{k}$ is tangential to the field line, and is therefore parallel to the vector $\mathbf{F} = F_x\mathbf{i} + F_y\mathbf{j} + F_z\mathbf{k}$. It follows that the differential equations of the field lines are

$$\frac{dx}{F_x} = \frac{dy}{F_y} = \frac{dz}{F_z}. \tag{1.13}$$

Problem 1.8 Sketch the two-dimensional fields: (i) $\mathbf{F} = \frac{1}{4}(x\mathbf{i} - y\mathbf{j})$, (ii) $\mathbf{G} = (x^2 + y^2)^{-\frac{1}{2}}(-y\mathbf{i} + x\mathbf{j})$, $(x, y) \neq (0, 0)$.

Solution. (i) According to the first of (1.13), the differential equation of the field lines is $dx/x = dy/(-y)$, giving

$$\ln x = -\ln y + C_1, \quad \text{or} \quad xy = C,$$

where C_1 and C are constants. Therefore the field lines are rectangular hyperbolae.

We note that $|\mathbf{F}| = \frac{1}{4}\sqrt{(x^2+y^2)}$, which is one-quarter of the distance of the point (x, y) from the origin, and also that F_x has the same sign as x. These considerations lead to Fig. (1.1).

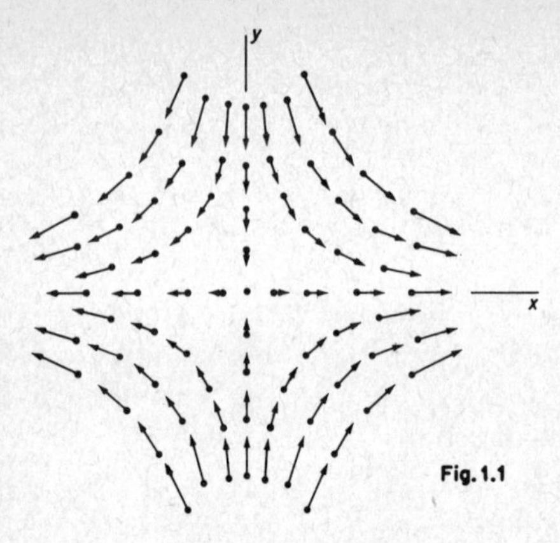

Fig. 1.1

(ii) Here the differential equation of the field lines reduces to

$$\frac{dx}{-y} = \frac{dy}{x},$$

i.e. $x\,dx + y\,dy = 0$, which integrates to give

$$x^2 + y^2 = C,$$

representing (for $C > 0$) circles centred at the origin. Furthermore, we have that $|\mathbf{G}| = 1$ and that G_x has the opposite sign to y. Hence we obtain Fig. 1.2.

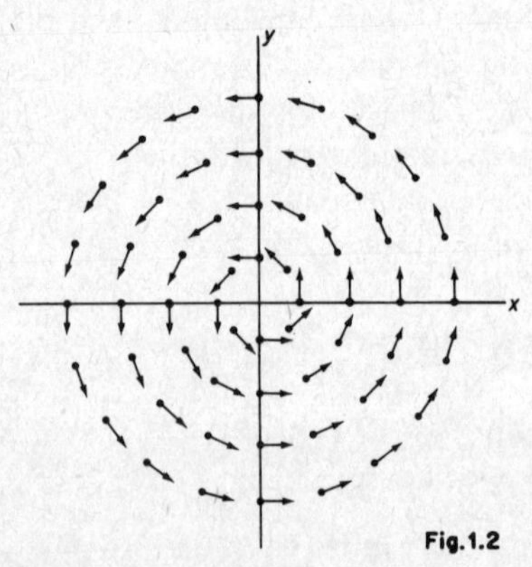

Fig. 1.2

□

Problem 1.9 Find the equations of the field lines for the vector fields: (i) $\mathbf{F} = -y\mathbf{i} + 4x\mathbf{j} + 2\mathbf{k}$, (ii) $\mathbf{G} = (y-z)\mathbf{i} + (z-x)\mathbf{j} + (x-y)\mathbf{k}$.

Solution. Equations (1.13) are

$$\frac{dx}{-y} = \frac{dy}{4x} = \frac{dz}{2}. \tag{1.14}$$

From the first pair, $4x\,dx + y\,dy = 0$, giving

$$4x^2 + y^2 = a^2 \tag{1.15}$$

as an integral, where a is a constant. Using this integral to obtain a second, we substitute $y = \pm\sqrt{(a^2 - 4x^2)}$, to get from (1.14)

$$dz = \pm 2\,dx/\sqrt{(a^2 - 4x^2)}.$$

Integrating, $$z = \pm \sin^{-1}(2x/a) + b,$$

i.e. $$z = \pm \sin^{-1}[2x/\sqrt{(y^2 + 4x^2)}] + b, \tag{1.16}$$

where b is a constant. This is a second integral.

Each of the integrals (1.15), (1.16) represents a family of surfaces. Taken simultaneously they represent the field lines as the lines of intersection of the first family with the second.

Given a ($\neq 0$), (1.15) is the equation of an elliptic cylinder with Oz as axis, meeting the xy-plane in the ellipse $4x^2 + y^2 = a^2$. Since $\mathbf{F}$ has a constant z component, it is evident that the field lines are spiral curves on such cylinders.

(ii) Here we must find two independent integrals of the differential equations

$$\frac{dx}{y-z} = \frac{dy}{z-x} = \frac{dz}{x-y}. \tag{1.17}$$

Denoting each of these common ratios by R, we have also that

$$\frac{l\,dx + m\,dy + n\,dz}{l(y-z) + m(z-x) + n(x-y)} = R, \tag{1.18}$$

for arbitrary multipliers l, m, n. Taking $l = m = n = 1$, we find that the denominator vanishes, and since the ratios (1.17) are not infinite for *all* values of x, y, z, it follows that the numerator must also vanish, i.e. $dx + dy + dz = 0$, which gives rise to the integral

$$x + y + z = a. \tag{1.19}$$

Again, taking $l = x$, $m = y$, $n = z$ makes the denominator of (1.18) vanish, so that for R to be (in general) finite we require $x\,dx + y\,dy + z\,dz = 0$. Hence we obtain a second integral

$$x^2 + y^2 + z^2 = b^2. \tag{1.20}$$

Equations (1.19), (1.20) represent the field lines as the intersections of parallel planes (with common normal direction (1, 1, 1)) and spheres centred at the origin. The field lines are therefore circles with centres on the line $x = y = z$, the planes of the circles being perpendicular to this line. □

Problem 1.10 Find the field lines for $\mathbf{F} = (1-y)\mathbf{i}+(1-x)\mathbf{j}+z\mathbf{k}$.

Solution. Equations (1.13) are

$$\frac{dx}{1-y} = \frac{dy}{1-x} = \frac{dz}{z}.$$

By multipliers,

$$\frac{dz}{z} = \frac{dx-dy}{x-y} = \frac{dx+dy}{2-(x+y)},$$

giving the integrals

$$\ln z = \ln(x-y)+a_1 = -\ln(x+y-2)+b_1,$$

or

$$z = a(x-y) = b(x+y-2)^{-1}, \tag{1.21}$$

where a_1, b_1, a and b are constants. Equations (1.21) represent the field lines. □

EXERCISES

1. In what regions are the following scalar and vector fields defined: (i) $\phi = \ln xyz$, (ii) $\mathbf{F} = \sin^{-1}(x-y)\mathbf{i}+[\ln(x^2+y^2)](\mathbf{j}+y\mathbf{k})$? Describe the regions geometrically.

2. Evaluate (i) grad x^2yz at the point (2, 3, 1), (ii) grad r^{-1}, ($r \neq 0$), where $r = \sqrt{(x^2+y^2+z^2)}$.

3. Find the rate of change of x^2+2xyz with respect to distance, in the direction of $\mathbf{i}-2\mathbf{j}-2\mathbf{k}$, at the point (1, 1, 4).

4. Show that the locus of points where the level surfaces of $\phi = x^2+y^2-z^2$ and $\psi = xyz$ cut orthogonally is the set of coordinate planes $x = 0$, $y = 0$ and $z = 0$.

5. The temperature in a region of space is $\phi = C \sin \pi(x^2+y^2-z)$, where C is a positive constant. An insect in the region is flying in the xy-plane and passes the point (1, 2, 0). Show that for any given speed of flight it will encounter the most rapid increase of temperature if the direction

of flight is $-(\mathbf{i}+2\mathbf{j})$. What rate of increase of temperature with respect to distance will it encounter in this case?

6. Sketch the vector field $\mathbf{F} = (x^2+y^2)^{-\frac{1}{2}}(-y\mathbf{i}+x\mathbf{j}+z\mathbf{k})$, $(x, y) \neq (0, 0)$. (*Hint*: Consider the behaviour of $\mathbf{F}$ on any cylinder $x^2+y^2 = \text{constant}$, using Problem 1.8 (ii) for $z = 0$.)

7. Find the equations of the field lines of the following vector fields:
 (i) $\mathbf{F} = (y^2-z^2)\mathbf{i}+(z^2-x^2)\mathbf{j}+(x^2-y^2)\mathbf{k}$,
 (ii) $\mathbf{G} = (y+z)\mathbf{i}+(z+x)\mathbf{j}+2z\mathbf{k}$.

Chapter 2

Integration of Scalar and Vector Fields

2.1 Line Integrals Let C be a curve

$$x = x(t), \quad y = y(t), \quad z = z(t). \tag{2.1}$$

where t takes all values in an interval $t_0 < t < t_1$, and the end-points A and B correspond to $t = t_0$ and $t = t_1$ respectively. If A and B coincide, C is said to be a *closed* curve. A curve which does not meet or cross itself (except, in the case of a closed curve, at the end-points) is *simple*.

Any scalar function $\phi(x, y, z)$ may be regarded as a function of t along C, by virtue of (2.1). The integral

$$\int_C \phi \, dt = \int_{t_0}^{t_1} \phi[x(t), y(t), z(t)] \, dt \tag{2.2}$$

(when it exists) is called the *line integral* of ϕ with respect to t along the curve C. The line integral always exists if $\phi(x, y, z)$ and $x(t)$, $y(t)$, $z(t)$ are continuous functions.

Problem 2.1 Find the line integral of $\phi(x, y, z) = (x^2+y^2)\sqrt{z}$, (i) with respect to t, (ii) with respect to arc distance s, along the curve

$$C: \quad x = t\cos t, \quad y = t\sin t, \quad z = t^2, \quad 0 \leqslant t \leqslant 1.$$

Solution. (i)

$$\begin{aligned} \int_C \phi \, dt &= \int_0^1 [(t\cos t)^2 + (t\sin t)^2] t \, dt \\ &= \int_0^1 t^3 \, dt \\ &= \left| \tfrac{1}{4}t^4 \right|_0^1 = \tfrac{1}{4}. \end{aligned}$$

(ii) The required line integral is, by definition,

$$\int_C \phi \, ds = \int_C \phi \frac{ds}{dt} dt, \tag{2.3}$$

where ds/dt may be found in terms of t as follows. Let

$$\mathbf{r}(t) = x(t)\mathbf{i} + y(t)\mathbf{j} + z(t)\mathbf{k}$$

denote the position vector of the point P on C, such that the arc distance along C from the initial point A $(t = 0)$ to P is s. Let P' be a point near P on C, such that the arc distance AP' is $s+\Delta s$. If $\mathbf{PP}' = \Delta\mathbf{r}$, then the vector

$$\frac{d\mathbf{r}}{ds} = \lim_{\Delta s \to 0} \frac{\Delta \mathbf{r}}{\Delta s} = \lim_{\Delta s \to 0} \frac{\mathbf{PP}'}{\Delta s}$$

is a *unit* vector in the direction of the tangent to C at P, since $|\Delta \mathbf{r}|/\Delta s \to 1$ as $\Delta s \to 0$. Taking magnitudes in the equation

$$\frac{d\mathbf{r}}{dt} = \frac{d\mathbf{r}}{ds}\frac{ds}{dt}$$

we get, therefore,

$$\begin{aligned} ds/dt = |d\mathbf{r}/dt| &= |x'(t)\mathbf{i}+y'(t)\mathbf{j}+z'(t)\mathbf{k}| \\ &= \sqrt{[x'^2(t)+y'^2(t)+z'^2(t)]} \\ &= \sqrt{[(\cos t - t\sin t)^2+(\sin t+t\cos t)^2+(2t)^2]} \\ &= \sqrt{(1+5t^2)}. \end{aligned}$$

Hence, by (2.3),

$$\begin{aligned} \int_C \phi\, ds &= \int_0^1 t^3\sqrt{(1+5t^2)}\, dt \\ &= \tfrac{1}{5}\int_0^1 [(1+5t^2)-1]\sqrt{(1+5t^2)}t\, dt \\ &= \tfrac{1}{5}\int_0^1 [(1+5t^2)^{\frac{3}{2}}-(1+5t^2)^{\frac{1}{2}}]t\, dt \\ &= \tfrac{1}{5}\left|\tfrac{1}{25}(1+5t^2)^{\frac{5}{2}}-\tfrac{1}{15}(1+5t^2)^{\frac{3}{2}}\right|_0^1 \\ &= 2(39\sqrt{6}+1)/375. \end{aligned}$$

□

Problem 2.2 If $\phi = x^2y(1+z)$, evaluate $\int_C \phi\, d\mathbf{r}$ where C is the curve $x = t$, $y = t^2$, $z = 1-t$, $\quad -1 \leqslant t \leqslant 1$.

Solution. On C, $\phi = t^4(2-t)$, and

$$dx = dt, \qquad dy = 2t\, dt, \qquad dz = -dt.$$

Hence,

$$\begin{aligned} \int_C \phi\, d\mathbf{r} &= \int_C \phi(dx\,\mathbf{i}+dy\,\mathbf{j}+dz\,\mathbf{k}) \\ &= \int_{-1}^1 t^4(2-t)(\mathbf{i}+2t\mathbf{j}-\mathbf{k})dt \\ &= \mathbf{i}\int_{-1}^1 (2t^4-t^5)\, dt+2\mathbf{j}\int_{-1}^1 (2t^5-t^6)\, dt-\mathbf{k}\int_{-1}^1 (2t^4-t^5)\, dt \\ &= 4(7\mathbf{i}-5\mathbf{j}-7\mathbf{k})/35. \end{aligned}$$

□

Problem 2.3 The electric field at the point $\mathbf{r} = x\mathbf{i}+y\mathbf{j}+z\mathbf{k}$ in a vacuum region of space is given by

$$\mathbf{E} = a\left(\frac{3x\mathbf{r}}{r^5}-\frac{\mathbf{i}}{r^3}\right), \tag{2.4}$$

where a is a constant. A wire loop in the region forms the circle $x = 2\cos t$, $y = 2\sin t$, $z = 0$, $0 \leqslant t \leqslant 2\pi$, and carries electric charge, the density (charge per unit length) being ρ. (i) Express the resultant force exerted by the field on the loop as a line integral. (ii) Calculate this force if ρ is a constant. (iii) Find the total moment of forces about the origin in this case.

Solution. By definition, $\mathbf{E}(x, y, z)$ denotes the force exerted by the field on a unit of charge placed at the point (x, y, z). Therefore an element of the loop with arc length Δs, carrying charge $\rho\,\Delta s$, is subject to a force $\mathbf{E}\rho\,\Delta s$. By dividing the whole loop into elements, summing, and taking the limit as all Δs approach zero, we arrive at the formula:

$$\text{resultant force} = \oint_C \mathbf{E}\rho\,ds = \mathbf{i}\oint_C E_x\rho\,ds + \mathbf{j}\oint_C E_y\rho\,ds + \mathbf{k}\oint_C E_z\rho\,ds, \quad (2.5)$$

where C is the circle formed by the wire (and a ring in an integral sign is used to emphasize that C is a closed curve). Substituting for x and $\mathbf{r}$ in terms of t, in (2.4), and using the fact that

$$\frac{ds}{dt} = \left|\frac{d\mathbf{r}}{dt}\right| = \sqrt{[(-2\sin t)^2 + (2\cos t)^2]} = 2, \quad (2.6)$$

we find that (2.5) reduces to

$$\text{resultant force} = \mathbf{F} = \tfrac{1}{4}a\int_0^{2\pi}[3\cos^2 t - 1)\mathbf{i} + 3\cos t\sin t\mathbf{j}]\rho\,dt.$$

(ii) If $\rho =$ constant, the last equation gives, on performing the integration, $\mathbf{F} = \frac{1}{4}\pi a\rho\mathbf{i}$.

(iii) The moment about O of the force $\mathbf{E}\rho\,\Delta s$ exerted on an element of wire of arc length Δs situated at the point $\mathbf{r}$ is $\Delta\mathbf{M} = \mathbf{r}\wedge\mathbf{E}\rho\,\Delta s$, and by summing over all elements, and taking the limit as all Δs approach zero, we obtain for the total moment of forces about O

$$\mathbf{M} = \lim_{\text{all }\Delta s\to 0}\sum_{\Delta s}\Delta\mathbf{M} = \oint_C(\mathbf{r}\wedge\mathbf{E})\rho\,ds. \quad (2.7)$$

But, on C,

$$\mathbf{r}\wedge\mathbf{E} = \tfrac{1}{8}a\begin{vmatrix}\mathbf{i} & \mathbf{j} & \mathbf{k}\\ 2\cos t & 2\sin t & 0\\ 3\cos^2 t - 1 & 3\cos t\sin t & 0\end{vmatrix} = \tfrac{1}{4}a\sin t\,\mathbf{k},$$

and since ρ is constant we get by (2.6), (2.7),

$$\mathbf{M} = \frac{a\rho}{4}\int_0^{2\pi}(\mathbf{r}\wedge\mathbf{E})\frac{ds}{dt}\,dt = \frac{a\rho}{2}\int_0^{2\pi}\sin t\,dt\,\mathbf{k} = 0. \qquad \square$$

Problem 2.4 Express as a line integral the work done by a force $\mathbf{F}(x, y, z)$ acting on a particle which describes a given curve C.

Solution. Let $\hat{\mathbf{t}}$ denote the unit tangent vector to C, in the sense s increasing, where as usual s denotes the arc distance measured along C from the initial point. The component of $\mathbf{F}$ in the forward tangential direction is $\mathbf{F}.\hat{\mathbf{t}}$, and therefore the work done by the force during an elementary displacement Δs along the curve is approximately $\mathbf{F}.\hat{\mathbf{t}}\,\Delta s$. If the whole path is divided into such elements, the total work done may be expressed as approximately

$$\sum_{\text{all } \Delta s} \mathbf{F}.\hat{\mathbf{t}}\,\Delta s,$$

and by taking the limit as all Δs approach zero we arrive at the (exact) formula for the total work done:

$$W = \int_C \mathbf{F}.\hat{\mathbf{t}}\,ds. \tag{2.8}$$

Write $\mathbf{r} = x\mathbf{i}+y\mathbf{j}+z\mathbf{k}$ for the position vector of a general point on C. Then, $d\mathbf{r} = \hat{\mathbf{t}}\,ds$, and (2.8) becomes

$$W = \int_C \mathbf{F}.d\mathbf{r} = \int_C (F_x\,dx+F_y\,dy+F_z\,dz). \tag{2.9}$$

A line integral of this form, where $\mathbf{F}$ is any given vector, is called the *tangential line integral* of $\mathbf{F}$ along C. □

Problem 2.5 Evaluate $\int_C \mathbf{F}.d\mathbf{r}$, where $\mathbf{F} = yz\mathbf{i}+2y\mathbf{j}-x^2\mathbf{k}$, and (i) C is the curve $x = t$, $y = t^2$, $z = t^3$, $0 \leqslant t \leqslant 1$; (ii) C consists of straight line segments from $(0,0,0)$ to the point $A(0,0,1)$, from A to $B(0,-3,1)$, and from B to $P(2,-3,1)$.

Solution. (i) by (2.9),

$$\int_C \mathbf{F}.d\mathbf{r} = \int_C (yz\,dx+2y\,dy-x^2\,dz).$$

By the equations defining C, $dx = dt$, $dy = 2t\,dt$, $dz = 3t^2\,dt$, and the line integral becomes on substitution for x, y and z,

$$\int_0^1 [t^5+(2t^2)(2t)-(t^2)(3t^2)]dt = \int_0^1 (t^5-3t^4+4t^3)\,dt = 17/30.$$

(ii) Along the line segment OA, $x = y = 0$, $dx = dy = 0$, and so, since $F_z = -x^2 = 0$,

$$\int_{OA} \mathbf{F}.d\mathbf{r} = \int_{OA} F_z\,dz = 0. \tag{2.10}$$

Along AB, $x = 0$, $z = 1$, $dx = dz = 0$, and

$$\int_{AB} \mathbf{F}.d\mathbf{r} = \int_{AB} F_y\,dy = \int_0^{-3} (2y)\,dy = \left| y^2 \right|_0^{-3} = 9. \tag{2.11}$$

Along BP, $y = -3, z = 1, dy = dz = 0$,

$$\int_{BP} \mathbf{F}\,.\,d\mathbf{r} = \int_{BP} F_x\,dx = \int_0^2 (yz)\,dx = \int_0^2 (-3)\,dx = -6. \tag{2.12}$$

Adding the contributions (2.10), (2.11), (2.12),

$$\int_C \mathbf{F}\,.\,d\mathbf{r} = 0+9-6 = 3. \qquad \square$$

Problem 2.6 The formula

$$\mathbf{B} = \frac{\mu I}{4\pi}\int_C \frac{d\mathbf{r}\wedge\mathbf{r}'}{r'^3} \tag{2.13}$$

gives the magnetic flux density $\mathbf{B}$ due to a steady electric current I flowing in a circuit C. Here, $\mathbf{r}$ is the position vector of a general point P on C relative to the origin, $\mathbf{r}'$ is the displacement from P to the point where $\mathbf{B}$ is measured, and μ is a certain physical constant. If C is the parabola $x = 2u$, $y = u^2$, $z = 1$, $-\infty < u < \infty$, calculate $\mathbf{B}$ at the point $(0, 0, 3)$.

Solution. For a general point u on C,

$$\mathbf{r} = 2u\mathbf{i}+u^2\mathbf{j}+\mathbf{k}, \qquad d\mathbf{r} = (2\mathbf{i}+2u\mathbf{j})\,du \tag{2.14}$$

At $(0,0,3)$,

$$\mathbf{r}' = 3\mathbf{k}-\mathbf{r} = -2u\mathbf{i}-u^2\mathbf{j}+2\mathbf{k}, \tag{2.15}$$

$$r'^3 = (\mathbf{r}'\,.\,\mathbf{r}')^{\frac{3}{2}} = (u^4+4u^2+4)^{\frac{3}{2}} = (u^2+2)^3. \tag{2.16}$$

By (2.13), ..., (2.16), we have at $(0, 0, 3)$, $d\mathbf{r}\wedge\mathbf{r}' = 2(2u\mathbf{i}-2\mathbf{j}+u^2\mathbf{k})$, and

$$\mathbf{B} = \frac{\mu I}{2\pi}\int_{-\infty}^{\infty} \frac{2u\mathbf{i}-2\mathbf{j}+u^2\mathbf{k}}{(u^2+2)^3}\,du.$$

The x component vanishes since $2u(u^2+2)^{-3}$ is an odd function of u. The remaining integrations may be carried out using, for example, the substitution $u = \sqrt{2}\tan\theta$, and are found to give

$$\mathbf{B}(0, 0, 3) = -\mu I(3\mathbf{j}-\mathbf{k})/32\sqrt{2}. \qquad \square$$

Problem 2.7 (i) If $\mathbf{F} = (e^x z-2xy)\mathbf{i}+(1-x^2)\mathbf{j}+(e^x+z)\mathbf{k}$, find a single-valued function $\phi(x, y, z)$ such that $\mathbf{F} = \text{grad}\,\phi$. (ii) Hence evaluate $\int_C \mathbf{F}\,.\,d\mathbf{r}$, where C is any path from the point $A(0, 1, -1)$ to $B(2, 3, 0)$.

Solution. (i) If ϕ exists, it must satisfy

$$\partial\phi/\partial x = F_x = e^x z-2xy, \tag{2.17}$$

$$\partial\phi/\partial y = F_y = 1-x^2, \tag{2.18}$$

$$\partial\phi/\partial z = F_z = e^x+z. \tag{2.19}$$

Integrating (2.17) with respect to x, with y and z constant,

$$\phi = e^x z - x^2 y + f(y, z), \tag{2.20}$$

where f is an arbitrary function. Substituting into (2.18),

$$-x^2 + \partial f/\partial y = 1 - x^2,$$

so that $\partial f/\partial y = 1$, and integration with respect to y with z constant gives

$$f = y + g(z), \tag{2.21}$$

where g is to be determined. By (2.19), (2.20),

$$e^x + \partial f/\partial z = e^x + dg/dz = e^x + z,$$

giving $g = \frac{1}{2}z^2 + c$, where c is a constant. Hence, by (2.20), (2.21),

$$\phi = e^x z + (1 - x^2)y + \tfrac{1}{2}z^2,$$

to within the addition of an arbitrary constant. The success of the construction automatically proves the existence of ϕ.

(ii) By (1.7),
$$\int_C \mathbf{F} \,.\, d\mathbf{r} = \int_C \operatorname{grad} \phi \,.\, d\mathbf{r} = \int_C d\phi = |\phi|_A^B$$
$$= \left| e^x z + (1 - x^2)y + \tfrac{1}{2}z^2 \right|_{(0,1,-1)}^{(2,3,0)}$$
$$= (-9) - (\tfrac{1}{2}) = -19/2. \qquad \square$$

This example illustrates the important result that if a vector $\mathbf{F}$ is the gradient of a single-valued scalar $\phi(x, y, z)$ in a region R, then the tangential line integral $\int_A^B \mathbf{F} \,.\, d\mathbf{r}$ is independent of the path of integration (in R) from the given point A to the given point B. The value of the line integral will always be $\phi(B) - \phi(A)$. Conversely, whenever the line integral depends only on the end-points of the path and not on the path itself, the vector $\mathbf{F}$ is expressible as the gradient of a single-valued scalar in the region in question. The proof is to be found in standard texts.

In such cases, we say that the field $\mathbf{F}$ is *conservative*, and the scalar $-\phi$ (defined to within a constant of addition) is called the *scalar potential* of $\mathbf{F}$.

Let M, N, P, Q be four points which occur in that order around any closed curve C in a region R in which $\mathbf{F}$ is conservative. Then,

$$\oint \mathbf{F} \,.\, d\mathbf{r} = \int_{MNP} \mathbf{F} \,.\, d\mathbf{r} + \int_{PQM} \mathbf{F} \,.\, d\mathbf{r} = \int_{MNP} \mathbf{F} \,.\, d\mathbf{r} - \int_{MQP} \mathbf{F} \,.\, d\mathbf{r} = 0, \tag{2.22}$$

where each integration is performed over all or part of the curve C. Hence, a condition that typifies a conservative field in R is that the tangential line integral around every closed curve in R is zero.

Problem 2.8 Show that if $\mathbf{F}$ is a conservative field then $\nabla \wedge \mathbf{F} \equiv 0$, where ∇ denotes the operator

$$\nabla = \mathbf{i}\frac{\partial}{\partial x}+\mathbf{j}\frac{\partial}{\partial y}+\mathbf{k}\frac{\partial}{\partial z}, \tag{2.23}$$

it being assumed that the potential ϕ possesses continuous second partial derivatives.

Solution. We have

$$\begin{aligned}\nabla \wedge \mathbf{F} &= \left(\mathbf{i}\frac{\partial}{\partial x}+\mathbf{j}\frac{\partial}{\partial y}+\mathbf{k}\frac{\partial}{\partial z}\right)\wedge(F_x\mathbf{i}+F_y\mathbf{j}+F_z\mathbf{k}) \\ &= \mathbf{i}\left(\frac{\partial F_z}{\partial y}-\frac{\partial F_y}{\partial z}\right)+\mathbf{j}\left(\frac{\partial F_x}{\partial z}-\frac{\partial F_z}{\partial x}\right)+\mathbf{k}\left(\frac{\partial F_y}{\partial x}-\frac{\partial F_x}{\partial y}\right)\end{aligned} \tag{2.24}$$

since $\mathbf{i}\wedge\mathbf{i} = 0$, $\mathbf{i}\wedge\mathbf{j} = \mathbf{k}$, etc. But $\mathbf{F} = \text{grad}\,\phi$, so that $F_x = \partial\phi/\partial x$, $F_y = \partial\phi/\partial y$, $F_z = \partial\phi/\partial z$. Hence (2.24) becomes

$$\nabla \wedge \mathbf{F} = \mathbf{i}\left(\frac{\partial^2\phi}{\partial y\partial z}-\frac{\partial^2\phi}{\partial z\partial y}\right)+\mathbf{j}\left(\frac{\partial^2\phi}{\partial z\partial x}-\frac{\partial^2\phi}{\partial x\partial z}\right)+\mathbf{k}\left(\frac{\partial^2\phi}{\partial x\partial y}-\frac{\partial^2\phi}{\partial y\partial x}\right) \equiv 0,$$

since $\partial^2\phi/\partial y\partial z = \partial^2\phi/\partial z\partial y$, etc., when both exist and are continuous. □

The converse result is: If $\nabla \wedge \mathbf{F} = 0$ everywhere in a simply-connected region R, then $\mathbf{F}$ is conservative in R. (A *simply-connected* region is one in which every closed curve can shrink continuously to zero without leaving the region.)

Note that in expanding the formal vector product (2.24) we keep the differentiation operators $\partial/\partial x$, $\partial/\partial y$, $\partial/\partial z$ to the left of the components of $\mathbf{F}$, and also that $\nabla\phi = \text{grad}\,\phi$. Further properties of the ∇ operator are considered in Chapter 3.

Problem 2.9 Test whether the following vector fields are conservative, and determine the potential where appropriate:

(i) $\mathbf{F} = yz\mathbf{i}+(x-y)\mathbf{j}+2xz\mathbf{k}$, (ii) $\mathbf{F} = (2x+yz)\mathbf{i}+(xz-2)\mathbf{j}+xy\mathbf{k}$.

Solution. (i) by (2.24),

$$\begin{aligned}\nabla \wedge \mathbf{F} &= \left[\frac{\partial}{\partial y}(2xz)-\frac{\partial}{\partial z}(x-y)\right]\mathbf{i}+\left[\frac{\partial}{\partial z}(yz)-\frac{\partial}{\partial x}(2xz)\right]\mathbf{j} \\ &\quad+\left[\frac{\partial}{\partial x}(x-y)-\frac{\partial}{\partial y}(yz)\right]\mathbf{k} \\ &= (0)\mathbf{i}+(y-2z)\mathbf{j}+(1-z)\mathbf{k},\end{aligned}$$

which is not identically zero. Therefore, $\mathbf{F}$ is not conservative and does not possess a scalar potential.

(ii) In this case,

$$\nabla\wedge\mathbf{F} = \left[\frac{\partial}{\partial y}(xy)-\frac{\partial}{\partial z}(xz-2)\right]\mathbf{i}+\left[\frac{\partial}{\partial z}(2x+yz)-\frac{\partial}{\partial x}(xy)\right]\mathbf{j}$$
$$+\left[\frac{\partial}{\partial x}(xz-2)-\frac{\partial}{\partial y}(2x+yz)\right]\mathbf{k}$$

$$= (0)\mathbf{i}+(0)\mathbf{j}+(0)\mathbf{k} = 0.$$

Hence $\mathbf{F}$ is conservative. The potential may be determined by the method of Problem 2.7. We find that $\mathbf{F} = \nabla(x^2+xyz-2y)$, whence the potential is $2y-xyz-x^2$, to within an arbitrary constant of addition. □

2.2 Surface Integrals A surface is a two-parameter set of points, given by equations of the form

$$x = x(u,v), \quad y = y(u,v), \quad z = z(u,v), \tag{2.25}$$

where u and v are the parameters. If the first pair of (2.25) can be solved for u and v, and we substitute the solution in the third equation, we obtain in place of (2.25) a single equation

$$z = f(x,y), \tag{2.26}$$

in which x and y act as the parameters. Other possible forms equivalent to (2.25) are $x = g(y,z)$ or $y = h(z,x)$. Yet another form of equation for a surface is (as we have seen) $\phi(x,y,z) = 0$, if we suppose that at least one of the variables x, y, z may be found, by solving, in terms of the other two.

Let ψ be a scalar which is defined at each point of the surface S. Divide S into n elements of area ΔS_i $(i = 1, 2, \ldots, n)$. If ψ_i denotes the value of ψ at an arbitrarily chosen point P_i in the ith element, we write

$$\int_S \psi\, dS = \lim_{n\to\infty}\sum_{i=1}^{n}\psi_i\,\Delta S_i, \tag{2.27}$$

provided that the limit on the right, formed as the linear dimensions of every ΔS_i tend to zero, exists and is independent of the mode of division and the choice of P_i. The integral (2.27) is called the *surface integral* of ψ over S. In the particular case when $\psi = 1$, we get

$$\int_S dS = \text{area of } S.$$

Problem 2.10 The surface S is given by $z = 1-x^2-y^2$, $z \geqslant 0$. (i) Find its area. (ii) If a thin film of matter is distributed on S so that the area mass density is $\rho(x,y) = a(x^2+2y^2)$ $(a = \text{const.})$, find the total mass on S.

Solution. The given surface is one of revolution about Oz, and is shown in Fig. 2.1. Let ΔS be a typical element intercepted by a rectangular column whose faces are parallel to one or other of the planes $x = 0$ or $y = 0$, the cross-sectional area of the column being $\Delta x\,\Delta y$. If $\hat{\mathbf{n}}$ is the unit normal (with positive z component) to S, we have by projection onto the xy-plane,

$$\Delta S\mathbf{k}\,.\,\hat{\mathbf{n}} = \Delta S\cos(\mathbf{k},\hat{\mathbf{n}}) = \Delta x\,\Delta y,$$

that is,

$$\Delta S = \Delta x\,\Delta y/(\mathbf{k}\,.\,\hat{\mathbf{n}}) \tag{2.28}$$

Hence the area of S is given by

$$\int_S dS = \iint_{S_0} \frac{dx\,dy}{\mathbf{k}\,.\,\hat{\mathbf{n}}}, \tag{2.29}$$

where S_0 is the projection of S on the xy-plane.

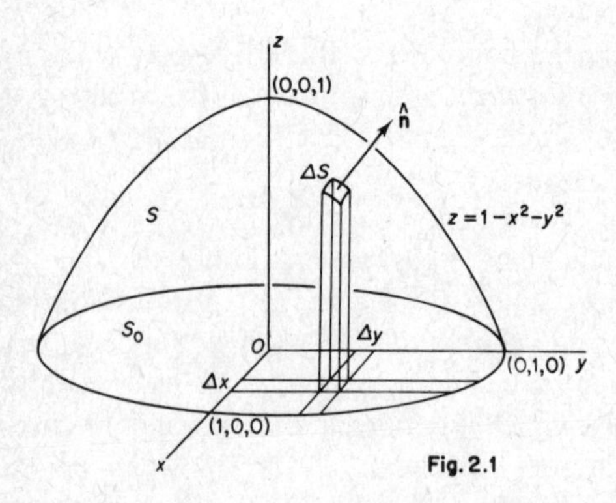

Fig. 2.1

Writing $\phi(x, y, z) \equiv 1 - x^2 - y^2 - z = 0$ for S, we have

$$\hat{\mathbf{n}} = -\frac{\nabla\phi}{|\nabla\phi|} = \frac{(2x\mathbf{i}+2y\mathbf{j}+\mathbf{k})}{\sqrt{(4x^2+4y^2+1)}},$$

where by inspection we have found it necessary to introduce a minus sign to obtain the unit normal with positive z component. Hence, forming $\mathbf{k}\,.\,\hat{\mathbf{n}}$ we obtain by (2.29)

$$\int_S dS = \iint_{S_0} \sqrt{(4x^2+4y^2+1)}\,dx\,dy. \tag{2.30}$$

This integral is most easily evaluated by transforming to polar coordinates R, ψ,

$$x = R\cos\psi, \qquad y = R\sin\psi, \tag{2.31}$$

the limits corresponding to S_0 being given by $0 \leqslant R \leqslant 1, 0 \leqslant \psi < 2\pi$.

It is necessary to replace $dx\,dy$ by $|J|\,dR\,d\psi$, in (2.30), where J is the Jacobian of the transformation, i.e.

$$J = \begin{vmatrix} \partial x/\partial R & \partial y/\partial R \\ \partial x/\partial \psi & \partial y/\partial \psi \end{vmatrix} = \begin{vmatrix} \cos\psi & \sin\psi \\ -R\sin\psi & R\cos\psi \end{vmatrix} = R.$$

Therefore,

$$\text{area} = \int_0^{2\pi}\int_0^1 \sqrt{(4R^2+1)}R\,dR\,d\psi = 2\pi\int_0^1 \sqrt{(4R^2+1)}R\,dR$$

$$= 2\pi\left|\tfrac{1}{12}(4R^2+1)^{\frac{3}{2}}\right| = \tfrac{1}{6}\pi(5\sqrt{5}-1).$$

(ii) The total mass is, by (2.29),

$$\int_S \rho\,dS = \iint_{S_0} a(x^2+2y^2)\frac{dx\,dy}{\mathbf{k}\,.\,\hat{\mathbf{n}}}$$

$$= a\iint_{S_0} (x^2+2y^2)\sqrt{(4x^2+4y^2+1)}\,dx\,dy$$

$$= \tfrac{1}{2}a\int_0^{2\pi}\int_0^1 (3-\cos 2\psi)\sqrt{(4R^2+1)}R^3\,dR\,d\psi \qquad \text{(polars)}$$

$$= 3a\pi\int_0^1 \sqrt{(4R^2+1)}R^3\,dR = (25\sqrt{5}+1)a\pi/40,$$

(using, for example, the substitution $2R = \sinh u$). □

Problem 2.11 The surface S has parametric equation $\mathbf{r} = \mathbf{r}(u, v)$, where u, v take all values in a region R of the uv-plane. Express the area of S as a double integral over R.

Solution. The lines $\mathbf{r} = \mathbf{r}(u, v_0)$ (on which v takes a fixed value v_0 as u varies) and $\mathbf{r} = \mathbf{r}(u_0, v)$ (on which u takes a fixed value u_0 as v varies) are *parametric lines* on S. As a typical area element ΔS we choose an elementary parallelogram formed by two pairs of neighbouring parametric lines, the vertices being the points $P(u_0, v_0)$, $Q(u_0+\Delta u, v_0)$, $T(u_0+\Delta u, v_0+\Delta v)$, $U(u_0, v_0+\Delta v)$ (Fig. 2.2). We have, to first order in Δu, Δv,

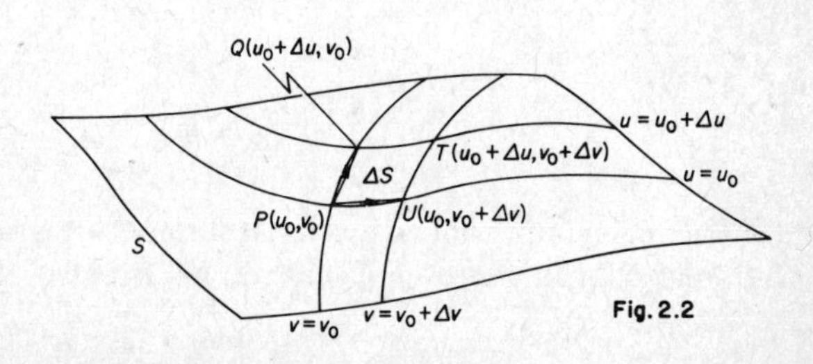

Fig. 2.2

$$\mathbf{PQ} = \mathbf{r}(u_0+\Delta u, v_0) - \mathbf{r}(u_0, v_0) = (\partial\mathbf{r}/\partial u)\Delta u,$$
$$\mathbf{PU} = \mathbf{r}(u_0, v_0+\Delta v) - \mathbf{r}(u_0, v_0) = (\partial\mathbf{r}/\partial v)\Delta v,$$

and hence to second order

$$\Delta S = PQ\,.\,PU \sin QPU = |\mathbf{PQ} \wedge \mathbf{PU}| = \left|\frac{\partial\mathbf{r}}{\partial u} \wedge \frac{\partial\mathbf{r}}{\partial v}\right| \Delta u\, \Delta v. \qquad (2.32)$$

By dividing the whole of S into such elements, summing, and forming the limit as Δu and Δv tend to zero we obtain the formula

$$\text{area} = \int_S dS = \iint_R \left|\frac{\partial\mathbf{r}}{\partial u} \wedge \frac{\partial\mathbf{r}}{\partial v}\right| du\, dv. \qquad (2.33)$$ □

Problem 2.12 Find the area of that part of the unit sphere

$$x = \sin\theta\cos\psi, \qquad y = \sin\theta\sin\psi, \qquad z = \cos\theta, \qquad (2.34)$$

for which $0 \leqslant \theta \leqslant \frac{1}{4}\pi$, $0 \leqslant \psi \leqslant \theta$.

Solution. We verify that by (2.34), $x^2+y^2+z^2 = 1$, which shows that the given surface is part of the unit sphere centred at O (θ and ψ are *spherical polar coordinates*; see next page. Putting $\mathbf{r} = x\mathbf{i}+y\mathbf{j}+z\mathbf{k}$, we have

$$\frac{\partial\mathbf{r}}{\partial\theta} = \frac{\partial x}{\partial\theta}\mathbf{i}+\frac{\partial y}{\partial\theta}\mathbf{j}+\frac{\partial z}{\partial\theta}\mathbf{k} = \cos\theta\cos\psi\,\mathbf{i}+\cos\theta\sin\psi\,\mathbf{j}-\sin\theta\,\mathbf{k},$$

$$\frac{\partial\mathbf{r}}{\partial\psi} = \frac{\partial x}{\partial\psi}\mathbf{i}+\frac{\partial y}{\partial\psi}\mathbf{j}+\frac{\partial z}{\partial\psi}\mathbf{k} = -\sin\theta\sin\psi\,\mathbf{i}+\sin\theta\cos\psi\,\mathbf{j}.$$

Forming the vector product as in (2.32), we find

$$\left|\frac{\partial\mathbf{r}}{\partial\theta} \wedge \frac{\partial\mathbf{r}}{\partial\psi}\right| = |\sin^2\theta\cos\psi\,\mathbf{i}+\sin^2\theta\sin\psi\,\mathbf{j}+\sin\theta\cos\theta\,\mathbf{k}|$$
$$= \sqrt{(\sin^4\theta+\sin^2\theta\cos^2\theta)} = \sin\theta,$$

so that the required area is

$$\int_0^{\frac{\pi}{4}}\int_0^{\theta} \left|\frac{\partial\mathbf{r}}{\partial\theta} \wedge \frac{\partial\mathbf{r}}{\partial\psi}\right| d\psi\, d\theta = \int_0^{\frac{\pi}{4}}\int_0^{\theta} \sin\theta\, d\psi\, d\theta$$
$$= \int_0^{\frac{\pi}{4}} \theta\sin\theta\, d\theta = \tfrac{1}{8}(4-\pi)\sqrt{2}.$$ □

Problems involving cylindrical and spherical surfaces are usually easier to solve using polar coordinates even if these do not occur in the original form of the problem.

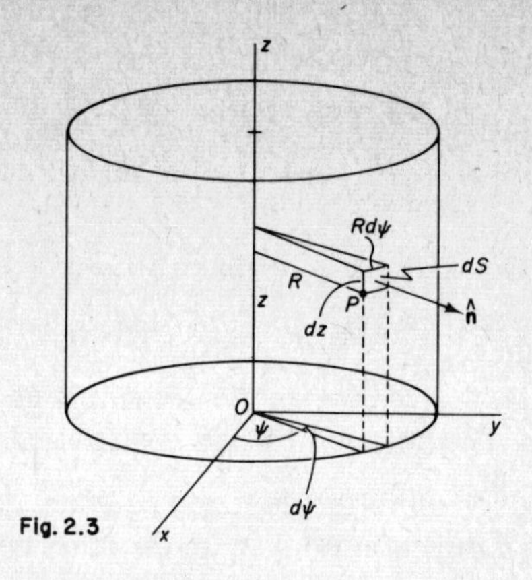

Fig. 2.3

Cylindrical polar coordinates (Fig. 2.3). P is the point (R, ψ, z):

$$x = R\cos\psi, \quad y = R\sin\psi, \quad z = z, \tag{2.35}$$
$$0 \leqslant R < \infty, \quad 0 \leqslant \psi < 2\pi, \quad -\infty < z < \infty.$$

Spherical polar coordinates (Fig. 2.4). P is the point (r, θ, ψ):

$$x = r\sin\theta\cos\psi, \quad y = r\sin\theta\sin\psi, \quad z = r\cos\theta, \tag{2.36}$$
$$0 \leqslant r < \infty, \quad 0 \leqslant \theta \leqslant \pi, \quad 0 \leqslant \psi < 2\pi.$$

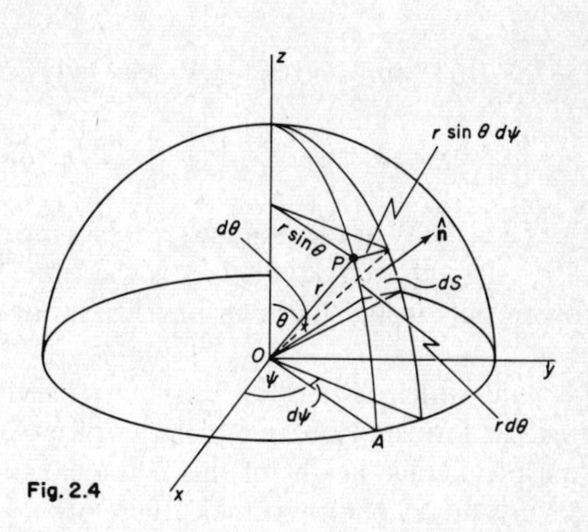

Fig. 2.4

Problem 2.13 Find expressions for the surface element dS and the outward-drawn unit normal $\hat{\mathbf{n}}$ in the appropriate polar coordinate system (i) for the cylinder $x^2 + y^2 = a^2$, (ii) for the sphere $x^2 + y^2 + z^2 = a^2$.

Solution. (i) By Fig. 2.3, $dS = R\,d\psi\,.\,dz = a\,d\psi\,dz$.

Resolving $\hat{\mathbf{n}}$,

$$\hat{\mathbf{n}} = \cos\psi\,\mathbf{i}+\sin\psi\,\mathbf{j}.$$

(ii) By Fig. 2.4, $dS = r\,d\theta\,.\,r\sin\theta\,d\psi = a^2\sin\theta\,d\theta\,d\psi$.
Resolving,

$$\hat{\mathbf{n}} = \sin\theta\,\widehat{\mathbf{OA}}+\cos\theta\,\mathbf{k} = \sin\theta(\cos\psi\,\mathbf{i}+\sin\psi\,\mathbf{j})+\cos\theta\,\mathbf{k}. \qquad \square$$

Surface integrals of the form $\int_S \mathbf{F}\,.\,\hat{\mathbf{n}}\,dS$, where $\hat{\mathbf{n}}$ is the unit normal to S in a specified sense, are known as *flux integrals* and are important in many applications. The notation $\int_S \mathbf{F}\,.\,d\mathbf{S}$ is also used, $d\mathbf{S} = \hat{\mathbf{n}}\,dS$ being a vector with magnitude dS in the direction $\hat{\mathbf{n}}$. When S is a *closed* surface, the outward normal direction is taken unless otherwise specified.

Problem 2.14 If $\mathbf{F} = y^2\mathbf{i}-y\mathbf{j}+xyz\,\mathbf{k}$, evaluate $\int_S \mathbf{F}\,.\,\hat{\mathbf{n}}\,dS$, where S is the curved surface of the cylinder $x^2+y^2 = 4$, $0 \leqslant z \leqslant 3$, and $\hat{\mathbf{n}}$ is directed away from the z-axis.

Solution. By Problem 2.13, $dS = 2\,d\psi\,dz$, $\hat{\mathbf{n}} = \cos\psi\,\mathbf{i}+\sin\psi\,\mathbf{j}$, and

$$\mathbf{F}\,.\,\hat{\mathbf{n}} = y^2\cos\psi-y\sin\psi = 2\sin^2\psi(2\cos\psi-1),$$

since $x = 2\cos\psi$, $y = 2\sin\psi$ on S. Therefore,

$$\int_S \mathbf{F}\,.\,\hat{\mathbf{n}}\,dS = \int_0^3\int_0^{2\pi} 2\sin^2\psi(2\cos\psi-1)\,.\,2\,d\psi\,dz$$

$$= 12\int_0^{2\pi}[2\sin^2\psi\cos\psi-\tfrac{1}{2}(1-\cos 2\psi)]\,d\psi$$

$$= 12\left|\tfrac{2}{3}\sin^3\psi-\tfrac{1}{2}(\psi-\tfrac{1}{2}\sin 2\psi)\right|_0^{2\pi} = -12\pi. \qquad \square$$

Problem 2.15 The velocity at the point (x, y, z) in a moving fluid is given by $\mathbf{v}(x, y, z, t) = (t-1)\mathbf{i}+z\mathbf{j}-x^2t^2\mathbf{k}$, where t is the time. Find the rate at which fluid is flowing out through the hemispherical surface $x^2+y^2+z^2 = 9$, $z \geqslant 0$, at $t = 1$.

Solution. The fluid which crosses an element of surface ΔS during a brief time interval Δt forms an oblique cylinder of cross-section ΔS and axis $\mathbf{v}\,\Delta t$. The perpendicular height of this cylinder is $(\mathbf{v}\,\Delta t)\,.\,\hat{\mathbf{n}}$, and the volume of fluid crossing ΔS per unit time is therefore

$$(\mathbf{v}\Delta t)\,.\,\hat{\mathbf{n}}\,\Delta S/\Delta t = \mathbf{v}\,.\,\hat{\mathbf{n}}\,\Delta S.$$

Hence the rate of volume flow across the whole of S is

$$\int_S \mathbf{v}\,.\,\hat{\mathbf{n}}\,dS. \qquad (2.37)$$

By (2.36), and Fig. 2.4, with $r = 3$,

$$\hat{\mathbf{n}} = 3(\sin\theta\cos\psi\,\mathbf{i}+\sin\theta\sin\psi\,\mathbf{j}+\cos\theta\,\mathbf{k}),$$
$$dS = 3\,d\theta\,.\,3\sin\theta\,d\psi = 9\sin\theta\,d\theta\,d\psi,$$
$$\mathbf{v} = z\mathbf{j}-x^2\mathbf{k} = 3\cos\theta\mathbf{j}-9\sin^2\theta\cos^2\psi\mathbf{k},$$

at $t = 1$. Therefore (2.37) becomes

$$\int_S 9\sin\theta\cos\theta(\sin\psi-3\sin\theta\cos^2\psi)\,dS$$
$$= 81\int_0^{\frac{1}{2}\pi}\int_0^{2\pi}\sin^2\theta\cos\theta(\sin\psi-3\sin\theta\cos^2\psi)\,d\psi\,d\theta$$
$$= -243\int_0^{\frac{1}{2}\pi}\sin^3\theta\cos\theta\left|\tfrac{1}{2}(\psi+\tfrac{1}{2}\sin 2\psi)\right|_0^{2\pi}d\theta$$
$$= -243\pi\left|\tfrac{1}{4}\sin^4\theta\right|_0^{\frac{1}{2}\pi} = -243\pi/4. \qquad \square$$

Problem 2.16 Find $\int_S \mathbf{r}\,.\,d\mathbf{S}$ and $\int_S \mathbf{F}\,.\,d\mathbf{S}$, where

$$\mathbf{r} = x\mathbf{i}+y\mathbf{j}+z\mathbf{k}, \quad \mathbf{F} = y\mathbf{i}-2x^2z\mathbf{k},$$

and S is the surface $z = 1-x^2-y^2$, $z \geqslant 0$ (with $d\mathbf{S}$ directed away from the xy-plane).

Solution. We can choose ΔS to have rectangular projection $\Delta x\,\Delta y$ on the xy-plane, where, by (2.28),

$$\mathbf{r}\,.\,\hat{\mathbf{n}}\,\Delta S = \frac{\mathbf{r}\,.\,\hat{\mathbf{n}}}{\mathbf{k}\,.\,\hat{\mathbf{n}}}\Delta x\,\Delta y,$$

and $\hat{\mathbf{n}}$ is the unit normal to S making an acute angle with $\mathbf{k}$. Hence,

$$\int_S \mathbf{r}\,.\,d\mathbf{S} = \iint_{S_0}\frac{\hat{\mathbf{r}}\,.\,\hat{\mathbf{n}}}{\mathbf{k}\,.\,\hat{\mathbf{n}}}\,dx\,dy = \iint_{S_0}\frac{\mathbf{r}\,.\,\mathbf{n}}{\mathbf{k}\,.\,\mathbf{n}}\,dx\,dy,$$

$\mathbf{n}$ being any normal vector to S, and S_0 being the projection of S on the xy-plane. Using a result in Problem 2.10, we take

$$\mathbf{n} = 2x\mathbf{i}+2y\mathbf{j}+\mathbf{k}, \qquad \mathbf{k}\,.\,\mathbf{n} = 1,$$
$$\mathbf{r}\,.\,\mathbf{n} = 2x^2+2y^2+z = 2x^2+2y^2+(1-x^2-y^2) = x^2+y^2+1$$

giving for the first of the required integrals

$$\iint_{x^2+y^2\leqslant 1}(x^2+y^2+1)\,dx\,dy = \int_0^{2\pi}\int_0^1(R^2+1)R\,dR\,d\psi = \tfrac{3}{2}\pi.$$

For the second integral,

$$\mathbf{F}.\mathbf{n} = (y\mathbf{i}-2x^2z\mathbf{k}).(2x\mathbf{i}+2y\mathbf{j}+\mathbf{k})$$
$$= 2xy-2x^2z = 2xy-2x^2(1-x^2-y^2).$$

Hence

$$\int_S \mathbf{F}.d\mathbf{S} = \iint_{S_0} (\mathbf{F}.\mathbf{n})/(\mathbf{k}.\mathbf{n})\,dx\,dy$$
$$= \iint_{x^2+y^2\leqslant 1} 2x[y+x(x^2+y^2-1)]\,dx\,dy$$
$$= 2\int_0^{2\pi}\int_0^1 R^2\cos\psi[\sin\psi+\cos\psi(R^2-1)]R\,dR\,d\psi$$
$$= 2\pi\int_0^1 R^3(R^2-1)\,dR = -\tfrac{1}{6}\pi. \qquad \square$$

Problem 2.17 Find $\int_S \mathbf{F}.d\mathbf{S}$, where $\mathbf{F} = x\mathbf{i}-2yz\mathbf{j}$ and S is the surface:

$$x = u\cos v,\quad y = u\sin v,\quad z = u^2, \tag{2.38}$$

where $0 \leqslant u \leqslant 1$, $0 \leqslant v < 2\pi$.

Solution. Writing $\mathbf{r} = x\mathbf{i}+y\mathbf{j}+z\mathbf{k}$, we have

$$\frac{\partial\mathbf{r}}{\partial u} = \cos v\,\mathbf{i}+\sin v\,\mathbf{j}+2u\mathbf{k},\quad \frac{\partial\mathbf{r}}{\partial v} = -u\sin v\,\mathbf{i}+u\cos v\,\mathbf{j},$$

whence

$$\frac{\partial\mathbf{r}}{\partial u}\wedge\frac{\partial\mathbf{r}}{\partial v} = -2u^2(\cos v\,\mathbf{i}+\sin v\,\mathbf{j})+u\mathbf{k}. \tag{2.39}$$

Since $\partial\mathbf{r}/\partial u$ and $\partial\mathbf{r}/\partial v$ are tangent vectors to the parametric lines $v =$ constant and $u =$ constant, respectively, on S, the vector (2.39) is normal to S. By (2.32), it follows that

$$d\mathbf{S} = \left(\frac{\partial\mathbf{r}}{\partial u}\wedge\frac{\partial\mathbf{r}}{\partial v}\right)du\,dv,$$

since the two sides agree in magnitude and direction. Hence, using (2.38),

$$\int_S \mathbf{F}.d\mathbf{S} = \int_0^{2\pi}\int_0^1 [x(-2u^2\cos v)-2yz(-2u^2\sin v)]\,du\,dv$$
$$= \int_0^{2\pi}\int_0^1 [-2u^3\cos^2 v+4u^5\sin^2 v]\,du\,dv = \tfrac{1}{6}\pi. \qquad \square$$

Problem 2.18 Express, as a surface integral, the solid angle subtended by a surface S at a point O.

Solution. Let ΔS be the area of an element of S containing a point P, whose position vector relative to O is $\mathbf{r}$, and let $\hat{\mathbf{n}}$ be a unit normal vector to S at P. The elementary cone formed by lines through O and boundary points of ΔS has cross-sectional area ΔA at P, where (approximately)

$$\Delta A = \Delta S|\cos(\hat{\mathbf{n}}, \hat{\mathbf{r}})| = \Delta S|\hat{\mathbf{n}} . \hat{\mathbf{r}}|,$$

the modulus sign being unnecessary if $\hat{\mathbf{n}}$ makes an acute angle with $\mathbf{r}$ as in Fig. 2.5. Let ΔA_0 be the area intercepted by the elementary cone on the

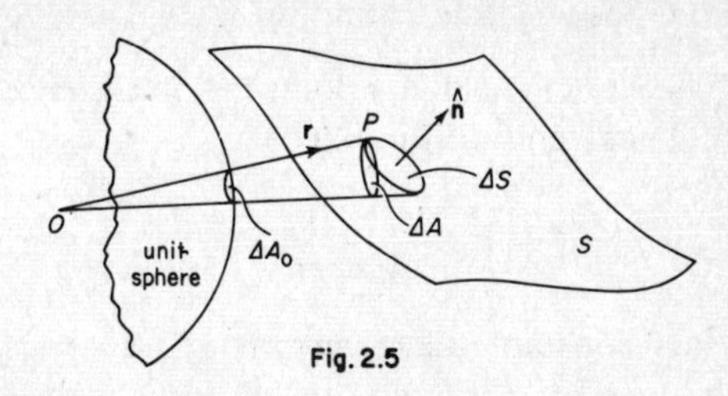

Fig. 2.5

unit sphere centered at O. By definition, the *magnitude* of the solid angle subtended at O by ΔS is

$$\Delta A_0 = \Delta A/OP^2 = \Delta S|\hat{\mathbf{n}} . \hat{\mathbf{r}}|/r^2.$$

When the sense of $\hat{\mathbf{n}}$ is prescribed, the solid angle subtended at O by the directed element $\hat{\mathbf{n}}\,\Delta S$ is (by definition) given in sign and magnitude by the expression $\Delta S\,\hat{\mathbf{n}} . \hat{\mathbf{r}}/r^2$. Thus, by summing, we have for the whole surface S:

$$\text{solid angle} = \int_S \frac{\hat{\mathbf{n}} . \hat{\mathbf{r}}}{r^2}\, dS = \int_S \frac{\mathbf{r} . d\mathbf{S}}{r^3}. \qquad (2.40)$$

If S is a *closed* surface (i.e. one which bounds a volume) the outward-drawn normal is taken in (2.40) unless otherwise specified. In this case the value of (2.40) is 4π if O is a point enclosed by S, and zero if O lies outside S. (Some surfaces, such as the Möbius strip, are *one-sided* or *non-orientable*. In such cases it is not possible to define a unique sense for $\hat{\mathbf{n}}$ over the whole surface, if the direction of $\hat{\mathbf{n}}$ is to vary continuously from point to point. See § 4.3.) □

Problem 2.19 Find the solid angle subtended at the origin by the disc S: $x^2+y^2 = a^2, z = b(b > 0)$, if $\hat{\mathbf{n}}$ points in the positive z direction.

Solution. Take plane polar coordinates (R, ψ) in the plane $z = b$. We may set $\hat{\mathbf{n}}\, dS = R\, dR\, d\psi\, \mathbf{k}$, the element dS being located at the point $\mathbf{r} = R\cos\psi\,\mathbf{i}+R\sin\psi\,\mathbf{j}+b\mathbf{k}$. Thus

$$\hat{\mathbf{n}} . \mathbf{r} = \mathbf{k} . \mathbf{r} = b, \quad r = (R^2+b^2)^{\frac{1}{2}},$$

so that (2.40) becomes

$$\int_S \frac{\mathbf{r}\,.\,\hat{\mathbf{n}}dS}{r^3} = \int_0^{2\pi}\int_0^a \frac{bR\,dR\,d\psi}{(R^2+b^2)^{\frac{3}{2}}}$$

$$= 2\pi b\left| -(R^2+b^2)^{-\frac{1}{2}}\right|_0^a = 2\pi[1-b(a^2+b^2)^{-\frac{1}{2}}],$$

which may also be written $2\pi(1-\cos\theta)$, where θ is the semi-angle of the cone of lines from O to the circle bounding S. □

Problem 2.20 The electric field at a point $P(\mathbf{r})$ *in vacuo* due to an electric point charge e at O is given (in SI units) by

$$\mathbf{E} = \frac{e}{4\pi\varepsilon_0}\frac{\hat{\mathbf{r}}}{r^2}, \tag{2.41}$$

where ε_0 is a physical constant (the *permittivity of free space*). Show that the flux of $\mathbf{E}$ through a surface S is $(e/4\pi\varepsilon_0)\omega$, where ω is the solid angle subtended by S at O.

Solution. By (2.40), (2.41), the required flux is

$$\int_S \mathbf{E}\,.\,d\mathbf{S} = \frac{e}{4\pi\varepsilon_0}\int_S \frac{\hat{\mathbf{r}}\,.\,d\mathbf{S}}{r^2} = \frac{e\omega}{4\pi\varepsilon_0},$$

Note that the flux through any closed surface enclosing O is e/ε_0. (This result is important in applications, and is a simple case of *Gauss's law*.) □

2.3 Volume Integrals Let $\phi(x, y, z)$ be a scalar field in a region V. Divide V into n elements of volume ΔV_i $(i = 1, 2, \ldots, n)$, and let ϕ_i denote the value of ϕ at any point (x_i, y_i, z_i) in the ith element. We write

$$\int_V \phi\,dV = \lim_{n\to\infty}\sum_{i=1}^{n}\phi_i\,\Delta V_i, \tag{2.42}$$

provided that the limit (as the division is made increasingly fine, so that the linear dimensions of all the ΔV_i approach zero) has a definite value which is independent of the mode of division. The integral is called the *volume integral* of ϕ over V.

If we take $\Delta V_i = \Delta x\,\Delta y\,\Delta z$, corresponding to a division of V into equal rectangular blocks (whose edges are all parallel to one or other of the coordinate axes), the limiting process leads to the expression of (2.42) as a triple (repeated) integral

$$\iiint \phi(x, y, z)\,dx\,dy\,dz \tag{2.43}$$

with limits appropriate to the region V.

If $\mathbf{F} = F_x\mathbf{i}+F_y\mathbf{j}+F_z\mathbf{k}$ is a vector field in V, then by definition,

$$\int_V \mathbf{F}\,dV = \left(\int_V F_x\,dV\right)\mathbf{i}+\left(\int_V F_y\,dV\right)\mathbf{j}+\left(\int_V F_z\,dV\right)\mathbf{k}, \tag{2.44}$$

if the separate integrals on the right exist.

Problem 2.21 Evaluate (2.44), where $\mathbf{F} = xy\mathbf{i}-zx\mathbf{j}+\mathbf{k}$, and V is the octant of the sphere $x^2+y^2+z^2 = 4$, $x \geqslant 0$, $y \geqslant 0$, $z \geqslant 0$.

Solution. Consider

$$\int_V F_x\,dV = \iiint_V xy\,dx\,dy\,dz.$$

Let the integration with respect to x be carried out first (corresponding to a summation over elementary blocks which form a prism with y and z constant, $0 \leqslant x \leqslant \sqrt{(4-y^2-z^2)}$). Next, let the integration with respect to y be performed (corresponding to a summation over prisms forming a slice with z constant, $0 \leqslant y \leqslant \sqrt{(4-z^2)}$). Finally, the integration with respect to z corresponds to a summation over all slices, $0 \leqslant z \leqslant 2$. Thus,

$$\begin{aligned}
\int_V F_x\,dV &= \int_0^2\int_0^{\sqrt{(4-z^2)}}\int_0^{\sqrt{(4-y^2-z^2)}} xy\,dx\,dy\,dz \\
&= \int_0^2\int_0^{\sqrt{(4-z^2)}} y\left|\tfrac{1}{2}x^2\right|_0^{\sqrt{(4-y^2-z^2)}} dy\,dz \\
&= \tfrac{1}{2}\int_0^2\int_0^{\sqrt{(4-z^2)}} y(4-y^2-z^2)\,dy\,dz \\
&= -\tfrac{1}{8}\int_0^2 \left|(4-y^2-z^2)^2\right|_{y=0}^{y=\sqrt{(4-z^2)}} dz \\
&= \tfrac{1}{8}\int_0^2 (4-z^2)^2\,dz = 32/15.
\end{aligned}$$

By symmetry,

$$\int_V F_y\,dV = -\int_V zx\,dV = -32/15.$$

Finally,

$$\int_V F_z\,dV = \int_V 1\,dV = V = 4\pi/3,$$

and so by (2.44),

$$\int_V \mathbf{F}\,dV = \frac{32}{15}(\mathbf{i}-\mathbf{j})+\frac{4\pi}{3}\mathbf{k}. \qquad \square$$

Problem 2.22 The velocity at the point (r, θ, ψ) (in spherical polar coordinates) in a fluid of constant density ρ, is

$$\mathbf{q} = [v(t)\cos\theta/r^3]\hat{\mathbf{r}}+[v(t)\sin\theta/2r^3]\hat{\boldsymbol{\theta}}, \tag{2.45}$$

where $\hat{\mathbf{r}}$, $\hat{\boldsymbol{\theta}}$ are unit vectors in the directions of r and θ increasing, respectively, t is the time and $v(t)$ is given. Find the momentum, at time t, of the fluid instantaneously in the region $0 \leqslant \theta \leqslant \frac{1}{3}\pi$ between the concentric spheres $r = 1$ and $r = 2$.

Solution. The mass of the fluid in a typical volume element ΔV is $\rho\,\Delta V$, and its momentum is approximately $\rho\mathbf{q}\Delta V$, where the value of $\mathbf{q}$ at any point in the element is taken. By summation, it follows that the instantaneous momentum of the fluid contained in a volume V is

$$\mathbf{M} = \int_V \rho\mathbf{q}\, dV = \iiint_V \rho\mathbf{q}\, dx\, dy\, dz. \tag{2.46}$$

Transforming to spherical polars, we replace $dx\,dy\,dz$ by $|J|dr\; d\theta\; d\psi$, where J is the Jacobian of the transformation (2.36):

$$J = \begin{vmatrix} \partial x/\partial r & \partial y/\partial r & \partial z/\partial r \\ \partial x/\partial \theta & \partial y/\partial \theta & \partial z/\partial \theta \\ \partial x/\partial \psi & \partial y/\partial \psi & \partial z/\partial \psi \end{vmatrix} = r^2 \sin\theta,$$

as is found on evaluating all the derivatives. Since ρ and $v(t)$ are independent of x, y and z, we get by (2.45), (2.46),

$$\mathbf{M} = \rho v(t) \int_0^{2\pi} \int_0^{\pi/3} \int_1^2 [(\cos\theta/r)\hat{\mathbf{r}} + (\sin\theta/2r)\hat{\boldsymbol{\theta}}] \sin\theta\; dr\; d\theta\; d\psi. \tag{2.47}$$

(Alternatively we can get $dV = r^2 \sin\theta\; dr\; d\theta\; d\psi$ by inspection of Fig. 5.2. See p. 59.) The vectors $\hat{\mathbf{r}}$ and $\hat{\boldsymbol{\theta}}$ are not constant throughout V and cannot be taken outside the integration. By Problem 2.11, these two vectors are in the directions of $\partial\mathbf{r}/\partial r$ and $\partial\mathbf{r}/\partial\theta$, respectively, and hence by (2.36),

$$\hat{\mathbf{r}} = \sin\theta\cos\psi\,\mathbf{i} + \sin\theta\sin\psi\,\mathbf{j} + \cos\theta\,\mathbf{k},$$

$$\hat{\boldsymbol{\theta}} = \cos\theta\cos\psi\,\mathbf{i} + \cos\theta\sin\psi\,\mathbf{j} - \sin\theta\,\mathbf{k}.$$

Substitution into (2.47) shows (if the integration with respect to ψ is carried out first) that only the $\mathbf{k}$ component survives, and

$$\mathbf{M} = \pi\rho\, v(t)\,\mathbf{k} \int_0^{\frac{\pi}{3}} \int_1^2 \frac{2\cos^2\theta - \sin^2\theta}{r} \sin\theta\; dr\; d\theta = \tfrac{3}{8}(\ln 2)\pi\rho\, v(t)\mathbf{k}.$$

Equation (2.45) has a physical interpretation; it represents the velocity $\mathbf{q}$ in an infinite, incompressible fluid in which a spherical body of unit radius is moving with speed $v(t)$. Axes are taken so that the centre of the sphere is instantaneously at the origin and moving along Oz, and the fluid velocity is zero at infinity. □

EXERCISES

1. If $\phi = (y+z)xyz$ and C is the curve $z = x$, $y = 2x^2$, $0 \leqslant x \leqslant 1$, evaluate (i) $\int_C \phi \, dx$, (ii) $\int_C \phi \, d\mathbf{r}$, where $\mathbf{r} = x\mathbf{i}+y\mathbf{j}+z\mathbf{k}$.

2. Evaluate $\int_C (x^2+y^2-z^2)\, ds$, where C is the curve $x = \sin t$, $y = t$, $z = \cos t$, $0 \leqslant t \leqslant \frac{1}{2}\pi$, and s denotes arc length.

3. The force acting on a particle located at the point (x, y, z) is $\mathbf{F} = xy\mathbf{i}+(y-z)\mathbf{j}+2x\mathbf{k}$. Find the work done by $\mathbf{F}$ when the particle moves along the curve $x = t$, $y = t^2$, $z = t^3$ from $t = 1$ to $t = 2$.

4. If $\mathbf{F} = (z\cos x+y^2)\mathbf{i}+2(x-2)y\mathbf{j}+\sin x\,\mathbf{k}$, and C is the path formed by straight line segments from $P(1, 0, 1)$ to O, from O to $Q(0, 2, 0)$, and from Q to $R(1, 2, 1)$, evaluate $\int_C \mathbf{F}\,.\,d\mathbf{r}$, (i) by performing the integration along each line segment in turn, (ii) by finding $\phi(x, y, z)$ such that $\mathbf{F} = \operatorname{grad}\phi$.

5. Find the area of the surface $(x-1)^2+y^2-2z = 0$, $0 \leqslant z \leqslant 2$.

6. Evaluate $\int_S \mathbf{F} \wedge d\mathbf{S}$, where $\mathbf{F} = z\mathbf{j}-\mathbf{k}$ and S is the spherical sector

$$x = \sin\theta\cos\psi,\ y = \sin\theta\sin\psi,\ z = \cos\theta, \quad 0 \leqslant \theta \leqslant \pi,\ 0 \leqslant \psi \leqslant \tfrac{1}{2}\pi.$$

7. Find the value of $\int_S (y\mathbf{i}+2x\mathbf{k})\,.\,d\mathbf{S}$, where S is the surface

$$x = \cos u\cos v,\quad y = \sin u\cos v,\quad z = \tfrac{1}{2}\sin v,\quad 0 \leqslant u \leqslant \tfrac{1}{2}\pi,\ 0 \leqslant v \leqslant \tfrac{1}{2}\pi.$$

8. Evaluate $\int_V (x+y+z)\, dV$, where V is the positive octant of the solid sphere $x^2+y^2+z^2 \leqslant 1$.

Chapter 3

The Operator Nabla. Formulae and Identities

3.1 Single Operations The operator $\nabla \equiv \mathbf{i}(\partial/\partial x)+\mathbf{j}(\partial/\partial y)+\mathbf{k}(\partial/\partial z)$, can act on a differentiable scalar function $\phi(x, y, z)$ or a differentiable vector function $\mathbf{F}(x, y, z) = F_x\mathbf{i}+F_y\mathbf{j}+F_z\mathbf{k}$, to give rise to three field quantities (of which the first has been discussed at length):

$$\text{(i)}\quad \operatorname{grad}\phi = \nabla\phi = \mathbf{i}\frac{\partial\phi}{\partial x}+\mathbf{j}\frac{\partial\phi}{\partial y}+\mathbf{k}\frac{\partial\phi}{\partial z};$$

$$\text{(ii)}\quad \operatorname{div}\mathbf{F} = \nabla.\mathbf{F} = \left(\mathbf{i}\frac{\partial}{\partial x}+\mathbf{j}\frac{\partial}{\partial y}+\mathbf{k}\frac{\partial}{\partial z}\right).(F_x\mathbf{i}+F_y\mathbf{j}+F_z\mathbf{k})$$

$$= \frac{\partial F_x}{\partial x}+\frac{\partial F_y}{\partial y}+\frac{\partial F_z}{\partial z};$$

$$\text{(iii)}\quad \operatorname{curl}\mathbf{F} = \nabla\wedge\mathbf{F} = \left(\mathbf{i}\frac{\partial}{\partial x}+\mathbf{j}\frac{\partial}{\partial y}+\mathbf{k}\frac{\partial}{\partial z}\right)\wedge(F_x\mathbf{i}+F_y\mathbf{j}+F_z\mathbf{k})$$

$$= \mathbf{i}\left(\frac{\partial F_z}{\partial y}-\frac{\partial F_y}{\partial z}\right)+\mathbf{j}\left(\frac{\partial F_x}{\partial z}-\frac{\partial F_z}{\partial x}\right)+\mathbf{k}\left(\frac{\partial F_y}{\partial x}-\frac{\partial F_x}{\partial y}\right)$$

$$= \begin{vmatrix} \mathbf{i} & \mathbf{j} & \mathbf{k} \\ \dfrac{\partial}{\partial x} & \dfrac{\partial}{\partial y} & \dfrac{\partial}{\partial z} \\ F_x & F_y & F_z \end{vmatrix}.$$

The second and third are known as the *divergence* and the *curl* of $\mathbf{F}$, respectively. The divergence is a scalar, while the curl is a vector; these quantities are not the same as $\mathbf{F}.\nabla$ and $-\mathbf{F}\wedge\nabla$ (which are both operators, since ∇ acts to the right) despite the formal resemblance to products of vectors.

Problem 3.1 If $\mathbf{m}$ is a constant vector, and $\mathbf{r} = x\mathbf{i}+y\mathbf{j}+z\mathbf{k}$, evaluate $\operatorname{grad}(\mathbf{m}.\mathbf{r})$.

Solution. Let $\mathbf{m} = m_x\mathbf{i}+m_y\mathbf{j}+m_z\mathbf{k}$. Then

$$\operatorname{grad}(\mathbf{m}.\mathbf{r}) = \nabla(\mathbf{m}.\mathbf{r}) = \left(\mathbf{i}\frac{\partial}{\partial x}+\mathbf{j}\frac{\partial}{\partial y}+\mathbf{k}\frac{\partial}{\partial z}\right)(m_x x+m_y y+m_z z).$$

But since m_x, m_y, m_z are constants, and x, y, z are independent variables,

$$\frac{\partial}{\partial x}(m_x x+m_y y+m_z z) = m_x \frac{\partial}{\partial x}(x) = m_x,$$

and by dealing with the remaining terms in like fashion we get

$$\text{grad}(\mathbf{m}.\mathbf{r}) = m_x\mathbf{i}+m_y\mathbf{j}+m_z\mathbf{k} = \mathbf{m}. \quad \square \quad (3.1)$$

Problem 3.2 Prove that

$$\nabla(uv) = u\nabla v+v\nabla u, \quad (3.2)$$

where $u = u(x, y, z)$ and $v = v(x, y, z)$ are differentiable functions.

Solution.

$$\begin{aligned}\nabla(uv) &= \mathbf{i}\frac{\partial}{\partial x}(uv)+\mathbf{j}\frac{\partial}{\partial y}(uv)+\mathbf{k}\frac{\partial}{\partial z}(uv)\\ &= \mathbf{i}\left(u\frac{\partial v}{\partial x}+v\frac{\partial u}{\partial x}\right)+\mathbf{j}\left(u\frac{\partial v}{\partial y}+v\frac{\partial u}{\partial y}\right)+\mathbf{k}\left(u\frac{\partial v}{\partial z}+v\frac{\partial u}{\partial z}\right)\\ &= u\left(\mathbf{i}\frac{\partial v}{\partial x}+\mathbf{j}\frac{\partial v}{\partial y}+\mathbf{k}\frac{\partial v}{\partial z}\right)+v\left(\mathbf{i}\frac{\partial u}{\partial x}+\mathbf{j}\frac{\partial u}{\partial y}+\mathbf{k}\frac{\partial u}{\partial z}\right)\\ &= u\,\nabla v+v\,\nabla u.\end{aligned} \quad \square$$

Problem 3.3 If $\mathbf{F} = 3xy\mathbf{i}+x^2z\mathbf{j}-y^2e^{2z}\mathbf{k}$, evaluate $\text{div}\,\mathbf{F}$ at the point $(1, 2, 0)$.

Solution.

$$\text{div}\,\mathbf{F} = \frac{\partial}{\partial x}(3xy)+\frac{\partial}{\partial y}(x^2z)+\frac{\partial}{\partial z}(-y^2e^{2z}) = 3y-2y^2e^{2z}.$$

At the point $(1, 2, 0)$, therefore, $\text{div}\,\mathbf{F} = -2$. $\square$

Problem 3.4 If $\mathbf{r} = x\mathbf{i}+y\mathbf{j}+z\mathbf{k}$, evaluate $\text{div}[f(r)\mathbf{r}]$, where f is any differentiable function. Hence show that the vector field $\mathbf{F} = r^{-2}\hat{\mathbf{r}}$ is solenoidal (i.e. has zero divergence).

Solution.

$$\text{div}[f(r)\mathbf{r}] = \frac{\partial}{\partial x}[f(r)x]+\frac{\partial}{\partial y}[f(r)y]+\frac{\partial}{\partial z}[f(r)z], \quad (3.3)$$

and

$$\frac{\partial r}{\partial x} = \frac{\partial}{\partial x}(x^2+y^2+z^2)^{\frac{1}{2}} = \tfrac{1}{2}(x^2+y^2+z^2)^{-\frac{1}{2}}2x = \frac{x}{r},$$

$$\frac{\partial r}{\partial y} = \frac{y}{r}, \quad \frac{\partial r}{\partial z} = \frac{z}{r}.$$

Hence the first term on the right in (3.3) is

$$\frac{\partial}{\partial x}[f(r)x] = \frac{df}{dr}\frac{\partial r}{\partial x}x+f = \frac{x^2}{r}\frac{df}{dr}+f,$$

and by writing similar expressions for the other two terms we get

$$\operatorname{div}[f(r)r] = \left(\frac{x^2+y^2+z^2}{r}\right)\frac{df}{dr}+3f = r\frac{df}{dr}+3f.$$

Putting $\mathbf{F} = r^{-2}\hat{\mathbf{r}} = r^{-3}\mathbf{r}$,

$$\operatorname{div}\mathbf{F} = r\frac{d}{dr}(r^{-3})+3r^{-3} = -3r^{-3}+3r^{-3} = 0.$$

Alternatively, one may apply the formula in the following problem to the vector $r^{-3}\mathbf{r}$, using the fact that

$$\operatorname{div}\mathbf{r} = \frac{\partial}{\partial x}(x)+\frac{\partial}{\partial y}(y)+\frac{\partial}{\partial z}(z) = 3. \qquad \square \qquad (3.4)$$

Problem 3.5 Show that

$$\nabla.(\phi\mathbf{F}) = \phi\nabla.\mathbf{F}+(\nabla\phi).\mathbf{F},$$

for differentiable functions ϕ and $\mathbf{F}$.

Solution.

$$\begin{aligned}\nabla.(\phi\mathbf{F}) &= \frac{\partial}{\partial x}(\phi F_x)+\frac{\partial}{\partial y}(\phi F_y)+\frac{\partial}{\partial z}(\phi F_z)\\ &= \left(\phi\frac{\partial F_x}{\partial x}+\frac{\partial\phi}{\partial x}F_x\right)+\left(\phi\frac{\partial F_y}{\partial y}+\frac{\partial\phi}{\partial y}F_y\right)+\left(\phi\frac{\partial F_z}{\partial z}+\frac{\partial\phi}{\partial z}F_z\right)\\ &= \phi\left(\frac{\partial F_x}{\partial x}+\frac{\partial F_y}{\partial y}+\frac{\partial F_z}{\partial z}\right)+\left(\frac{\partial\phi}{\partial x}F_x+\frac{\partial\phi}{\partial y}F_y+\frac{\partial\phi}{\partial z}F_z\right)\\ &= \phi\nabla.\mathbf{F}+(\nabla\phi).\mathbf{F}.\end{aligned} \qquad \square$$

Problem 3.6 Evaluate (i) $(\nabla.\mathbf{a})\mathbf{b}$, (ii) $(\mathbf{a}.\nabla)\mathbf{b}$, at the point $(-2, 1, 1)$, where $\mathbf{a} = xy\mathbf{i}-z\mathbf{j}+2yz\mathbf{k}$, $\mathbf{b} = 3z^2\mathbf{i}+xy\mathbf{j}+\mathbf{k}$.

Solution. (i)

$$\nabla.\mathbf{a} = \frac{\partial}{\partial x}(xy)+\frac{\partial}{\partial y}(-z)+\frac{\partial}{\partial z}(2yz) = 3y,$$

and therefore at $(-2,1,1)$,

$$(\nabla.\mathbf{a})\mathbf{b} = 3y(3z^2\mathbf{i}+xy\mathbf{j}+\mathbf{k}) = 9\mathbf{i}-6\mathbf{j}+3\mathbf{k}.$$

Problem 3.11 Prove that $\nabla \wedge (\mathbf{m} \wedge \mathbf{r}) = 2\mathbf{m}$, if $\mathbf{m}$ = constant.

Solution. Let $\mathbf{m} = m_x\mathbf{i}+m_y\mathbf{j}+m_z\mathbf{k}$. Then

$$\mathbf{m} \wedge \mathbf{r} = (m_y z - m_z y)\mathbf{i} + (m_z x - m_x z)\mathbf{j} + (m_x y - m_y x)\mathbf{k},$$

$$\nabla \wedge (\mathbf{m} \wedge \mathbf{r}) = \begin{vmatrix} \mathbf{i} & \mathbf{j} & \mathbf{k} \\ \partial/\partial x & \partial/\partial y & \partial/\partial z \\ m_y z - m_z y & m_z x - m_x z & m_x y - m_y x \end{vmatrix}.$$

The x component of this expression is

$$\frac{\partial}{\partial y}(m_x y - m_y x) - \frac{\partial}{\partial z}(m_z x - m_x z) = m_x - (-m_x) = 2m_x,$$

and treating the other components in the same way we get

$$\nabla \wedge (\mathbf{m} \wedge \mathbf{r}) = 2m_x\mathbf{i} + 2m_y\mathbf{j} + 2m_z\mathbf{k} = 2\mathbf{m}. \qquad \square$$

3.2 Successive Operations

Problem 3.12 Evaluate (i) $\nabla^2\phi$; (ii) $\nabla^2\mathbf{F}$, where $\phi = x^2y + z^2\cos y$ and $\mathbf{F} = 2y\mathbf{i} + x^2z^3\mathbf{j} - x^3\mathbf{k}$.

Solution. The operator ∇^2 is defined to be

$$\nabla^2 \equiv \nabla . \nabla = \left(\mathbf{i}\frac{\partial}{\partial x} + \mathbf{j}\frac{\partial}{\partial y} + \mathbf{k}\frac{\partial}{\partial z}\right) . \left(\mathbf{i}\frac{\partial}{\partial x} + \mathbf{j}\frac{\partial}{\partial y} + \mathbf{k}\frac{\partial}{\partial z}\right) = \frac{\partial^2}{\partial x^2} + \frac{\partial^2}{\partial y^2} + \frac{\partial^2}{\partial z^2}.$$

(i) $$\nabla^2\phi = \frac{\partial^2\phi}{\partial x^2} + \frac{\partial^2\phi}{\partial y^2} + \frac{\partial^2\phi}{\partial z^2} = 2y - z^2\cos y + 2\cos y.$$

(ii) $$\begin{aligned} \nabla^2\mathbf{F} &= \nabla^2(2y\mathbf{i} + x^2z^3\mathbf{j} - x^3\mathbf{k}) \\ &= \mathbf{i}\nabla^2(2y) + \mathbf{j}\nabla^2(x^2z^3) + \mathbf{k}\nabla^2(-x^3) \\ &= (2z^3 + 6x^2z)\mathbf{j} - 6x\mathbf{k}. \end{aligned} \qquad \square$$

Problem 3.13 (i) Find $\nabla^2 f(r)$, where f is a twice differentiable function. (ii) Find all values of the constant n such that $\nabla^2 r^n = 0$.

Solution. (i) By (3.6),

$$\begin{aligned} \nabla^2 f(r) &= \nabla . [\nabla f(r)] = \nabla . \left(\frac{df}{dr}\hat{\mathbf{r}}\right) = \nabla . \left(r^{-1}\frac{df}{dr}\mathbf{r}\right) \\ &= r^{-1}\frac{df}{dr}\nabla . \mathbf{r} + \nabla\left(r^{-1}\frac{df}{dr}\right) . \mathbf{r}, \end{aligned}$$

by Problem 3.5. But $\nabla . \mathbf{r} = 3$, and by applying (3.6) again we get

$$\nabla^2 f(r) = 3r^{-1}\frac{df}{dr} + \left(r^{-1}\frac{d^2f}{dr^2} - r^{-2}f\right)\hat{\mathbf{r}} . \mathbf{r}$$

$$= \frac{d^2f}{dr^2} + \frac{2}{r}f.$$

(ii) Putting $f = r^n$,

$$\nabla^2 r^n = n(n-1)r^{n-2} + 2nr^{n-2}$$
$$= n(n+1)r^{n-2} = 0,$$

when n takes either of the values 0 or -1.

Problem 3.14 If $\mathbf{F} = 2xz\mathbf{i} + y^2\mathbf{j} - xy\mathbf{k}$, show that $\nabla . (\nabla \wedge \mathbf{F}) = 0$.

Solution.

$$\nabla \wedge \mathbf{F} = \begin{vmatrix} \mathbf{i} & \mathbf{j} & \mathbf{k} \\ \partial/\partial x & \partial/\partial y & \partial/\partial z \\ 2xz & y^2 & -xy \end{vmatrix}$$

$$= \left[\frac{\partial}{\partial y}(-xy) - \frac{\partial}{\partial z}(y^2)\right]\mathbf{i} + \left[\frac{\partial}{\partial z}(2xz) - \frac{\partial}{\partial x}(-xy)\right]\mathbf{j}$$
$$+ \left[\frac{\partial}{\partial x}(y^2) - \frac{\partial}{\partial y}(2xz)\right]\mathbf{k}$$

$$= -x\mathbf{i} + (2x+y)\mathbf{j};$$

$$\nabla . (\nabla \wedge \mathbf{F}) = \frac{\partial}{\partial x}(-x) + \frac{\partial}{\partial y}(2x+y) = -1+1 = 0.$$

This is a particular case of a general result in the next problem. □

Problem 3.15 Prove that div curl $\mathbf{F} \equiv 0$, if $\mathbf{F}$ is continuous and has continuous partial derivatives of second order.

Solution. Expressing $\mathbf{F}$ in component form we find

$$\text{div curl } \mathbf{F} = \nabla . (\nabla \wedge \mathbf{F})$$
$$= \frac{\partial}{\partial x}\left(\frac{\partial F_z}{\partial y} - \frac{\partial F_y}{\partial z}\right) + \frac{\partial}{\partial y}\left(\frac{\partial F_x}{\partial z} - \frac{\partial F_z}{\partial x}\right) + \frac{\partial}{\partial z}\left(\frac{\partial F_y}{\partial x} - \frac{\partial F_x}{\partial y}\right)$$
$$= 0,$$

since $\partial^2 F_y/\partial x \partial z = \partial^2 F_y/\partial z \partial x$ etc. under the stated conditions on $\mathbf{F}$. □

Problem 3.16 If $\phi(x, y, z)$ is a solution of *Laplace's equation* $\nabla^2\phi = 0$, the level surfaces $\phi(x, y, z) =$ constant are said to be *equipotential surfaces.* Show that the condition for a given family of surfaces $f(x, y, z) =$ constant to be equipotential is that $\nabla^2 f/(\nabla f)^2$ be a function of f only (it being assumed that f is twice continuously differentiable). In this case, determine the potential ϕ in terms of f.

Solution. We first show the condition to be necessary, i.e. we assume that there is a solution of the equation $\nabla^2\phi = 0$ such that the level surfaces of ϕ coincide with the level surfaces of f. Thus, to each possible value of f there corresponds a value of ϕ, which means that ϕ may be regarded as a function of f alone, $\phi = F(f)$ say. (For example, the spheres $r =$ constant are equipotential surfaces since $\phi = r^{-1}$ is a solution of Laplace's equation. In this case we have $\phi = F(r) = r^{-1}$. It is *not* true that r is itself a solution of Laplace's equation, however.)

We have (with a prime denoting differentiation)

$$\begin{aligned}\nabla\phi &= \nabla F(f) = F'(f)\nabla f,\\ \nabla^2\phi &= \nabla.(\nabla F) = \nabla.[F'\nabla f]\\ &= F'\nabla.(\nabla f)+(\nabla F').(\nabla f)\\ &= F'\nabla^2 f + F''(\nabla f)^2 = 0, \qquad (3.8)\end{aligned}$$

(given), by Problem 3.5. We cannot, at a general point, have $(\nabla f)^2 = 0$, because this requires the vanishing of the partial derivatives of f with respect to x, y and z, which implies that f is not strictly dependent on these variables. Hence we may divide to get

$$\frac{\nabla^2 f}{(\nabla f)^2} = -\frac{F''}{F'}, \qquad (3.9)$$

and the right-hand side is evidently a function of f only, which proves that the stated condition is necessary.

Sufficiency follows from the fact that when the left-hand side of (3.9) is a function of f only, we can integrate to obtain $F(f)$ such that $\phi = F(f)$ is a solution of Laplace's equation (as follows by reversing the steps leading to (3.8)). Writing (3.9) as

$$\frac{d}{df}(\ln F') = -\frac{\nabla^2 f}{(\nabla f)^2},$$

we obtain after two integrations

$$\phi = F(f) = A\int \exp\left\{-\int \frac{\nabla^2 f}{(\nabla f)^2}\,df\right\}df + B, \qquad (3.10)$$

where A and B are arbitrary constants. This gives the potential ϕ explicitly. □

Problem 3.17 Determine whether the following families are equipotential surfaces, and give the potential function where appropriate:

(i) $y^2+z^2 = ce^x$, (ii) $x^3+y^3+z^3 = c$, (iii) $x^2-2cx+y^2+1 = 0$.

Solution. (i) Let f denote $(y^2+z^2)e^{-x}$. We find

$$\nabla f = e^{-x}[-(y^2+z^2)\mathbf{i}+2y\mathbf{j}+2z\mathbf{k}],$$
$$\nabla^2 f = e^{-x}(y^2+z^2+4),$$
$$\nabla^2 f/(\nabla f)^2 = e^x/(y^2+z^2) = 1/f.$$

Since the last expression is a function of f alone, the family is equipotential. We get (to within a constant which does not affect the final result)

$$\int \frac{\nabla^2 f}{(\nabla f)^2}\, df = \ln f,$$

and substitution in (3.10) then gives $\phi = A \ln f + B$.

(ii) Here we put $f = x^3+y^3+z^3$, and find

$$\frac{\nabla^2 f}{(\nabla f)^2} = \frac{2(x+y+z)}{3(x^4+y^4+z^4)},$$

which evidently cannot be expressed as a function of $x^3+y^3+z^3$ alone. Hence the family is not equipotential.

(iii) The family is given by $f = c$, where

$$x^2-2fx+y^2+1 = 0. \tag{3.11}$$

Instead of solving for f, we may treat f as an implicit function defined by (3.11). Differentiating partially with respect to x, with y and z constant, (a suffix denoting differentiation)

$$2x-2f_x x-2f = 0, \quad \text{i.e. } f_x = 1-x^{-1}f.$$

Differentiating again,

$$f_{xx} = x^{-2}f-x^{-1}f_x = 2x^{-2}f-x^{-1}.$$

Forming the other necessary derivatives we thus find

$$\nabla^2 f/(\nabla f)^2 = 2f/(f^2-1),$$

which shows that the family is equipotential. The potential function is found from (3.10) as in (i). It is

$$\phi = A_1 \ln\left(\frac{f+1}{f-1}\right)+B,$$

where $A_1\,(= -\frac{1}{2}A)$ is a constant. □

3.3 List of Identities The following identities are listed for reference. Those not proved in foregoing problems may be verified by expressing each side in component form. It is assumed that all relevant derivatives exist and are continuous.

(i) $\nabla(\phi\psi) = \phi\nabla\psi+\psi\nabla\phi$,
(ii) $\nabla\,.\,(\phi\mathbf{A}) = \phi\nabla\,.\,\mathbf{A}+(\nabla\phi)\,.\,\mathbf{A}$,
(iii) $\nabla\,.\,(\nabla\wedge\mathbf{A}) = 0$,
(iv) $\nabla\,.\,(\mathbf{A}\wedge\mathbf{B}) = \mathbf{B}\,.\,(\nabla\wedge\mathbf{A})-\mathbf{A}\,.\,(\nabla\wedge\mathbf{B})$,
(v) $\nabla\wedge(\phi\mathbf{A}) = \phi\nabla\wedge\mathbf{A}+(\nabla\phi)\wedge\mathbf{A}$,
(vi) $\nabla\wedge\nabla\phi = 0$,
(vii) $\nabla\wedge(\nabla\wedge\mathbf{F}) = \nabla(\nabla\,.\,\mathbf{F})-\nabla^2\mathbf{F}$,
(viii) $\nabla(\mathbf{A}\,.\,\mathbf{B}) = (\mathbf{A}\,.\,\nabla)\mathbf{B}+(\mathbf{B}\,.\,\nabla)\mathbf{A}+\mathbf{A}\wedge(\nabla\wedge\mathbf{B})+\mathbf{B}\wedge(\nabla\wedge\mathbf{A})$,
(ix) $\nabla\wedge(\mathbf{A}\wedge\mathbf{B}) = (\nabla\,.\,\mathbf{B})\mathbf{A}-(\nabla\,.\,\mathbf{A})\mathbf{B}+(\mathbf{B}\,.\,\nabla)\mathbf{A}-(\mathbf{A}\,.\,\nabla)\mathbf{B}$,
(x) $\mathbf{A}\wedge(\nabla\wedge\mathbf{A}) = \frac{1}{2}\nabla A^2-(\mathbf{A}\,.\,\nabla)\mathbf{A}$.

EXERCISES

1. Evaluate $\nabla\,.\,(xe^y\mathbf{i}-\sin xy\,\mathbf{j}+z\mathbf{k})$.

2. Evaluate $\nabla\wedge(y^2z\mathbf{i}+z^2x\mathbf{j}+x^2y\mathbf{k})$.

3. If $\mathbf{F} = \sin r\,\hat{\mathbf{r}}$, and $\mathbf{r} = x\mathbf{i}+y\mathbf{j}+z\mathbf{k}$, find (i) $\nabla\,.\,\mathbf{F}$, (ii) $\nabla\wedge\mathbf{F}$.

4. If $\phi = x^2yz^3$ and $\mathbf{F} = x^2\mathbf{i}+y^2\mathbf{j}+z^2\mathbf{k}$, evaluate (i) $\nabla^2\phi$, (ii) $\nabla^2\mathbf{F}$.

5. Verify the identity (§ 3.3(vii)):

$$\nabla\wedge(\nabla\wedge\mathbf{F}) = \nabla(\nabla\,.\,\mathbf{F})-\nabla^2\mathbf{F},$$

for the vector $\mathbf{F} = x^2y\mathbf{i}+zx\mathbf{j}-3yz^2\mathbf{k}$, by evaluating the two sides.

6. Show that

$$\nabla\,.\,[\phi(u,v)\nabla u\wedge\nabla v] = 0,$$

where u and v are functions of x, y and z, with continuous second partial derivatives, and ϕ is a continuously differentiable function of u and v.

7. Evaluate (i) $\nabla^2\ln(x^2+y^2)$, (ii) $\nabla^2\ln(x^2+y^2+z^2)$.

8. Show that the family of surfaces

$$(x-a)^2+y^2 = c[(x+a)^2+y^2]$$

are equipotentials, a being a constant and c a parameter for the family. Find the potential function.

Chapter 4

Integral Theorems

4.1 Green's Theorem in the *xy*-Plane This theorem states that if $P(x, y)$ and $Q(x, y)$ are continuous functions with continuous partial derivatives in a region R of the plane and its boundary, which is a simple closed curve C, then

$$\oint_C (P\,dx+Q\,dy) = \iint_R \left(\frac{\partial Q}{\partial x}-\frac{\partial P}{\partial y}\right) dx\,dy, \tag{4.1}$$

where C is described in the anticlockwise direction.

If R has the properties (i) that any line x = constant, which crosses R, meets C in just two points $y_1(x)$, $y_2(x)$, $(y_1 < y_2)$; (ii) that any line y = constant, which crosses R, meets C in just two points $x_1(y)$, $x_2(y)$, $(x_1 < x_2)$; the proof is quite easy. Let $x = a$, $x = b$ be the tangents to C parallel to Oy on the left- and right-hand sides of R respectively. Then

$$\begin{aligned}\oint_C P\,dx &= \int_a^b P[x, y_1(x)]\,dx + \int_b^a P[x, y_2(x)]\,dx \\ &= -\int_a^b \{P[x, y_2(x)] - P[x, y_1(x)]\}\,dx \\ &= -\int_a^b \left(\int_{y_1(x)}^{y_2(x)} (\partial P/\partial y)\,dy\right) dx \\ &= -\iint_R (\partial P/\partial y)\,dx\,dy. \end{aligned} \tag{4.2}$$

A similar calculation with the roles of x and y interchanged shows that

$$\oint_C Q\,dy = \iint_R (\partial Q/\partial x)\,dx\,dy, \tag{4.3}$$

and by adding (4.2), (4.3), the result follows.

The proof readily extends to a more general type of region R which can be dissected into a number, n, of subregions to each of which (i) and (ii) apply. For each subregion there is an identity of the form (4.1), and the required result follows on adding the n identities, since the contributions from the left-hand sides along the lines of dissection cancel in pairs.

If we denote the vector $P\mathbf{i}+Q\mathbf{j}$ by $\mathbf{F}$, we find that

$$\nabla \wedge \mathbf{F} = \left(\frac{\partial Q}{\partial x}-\frac{\partial P}{\partial y}\right)\mathbf{k},$$

and so by putting $d\mathbf{S} = dx\,dy\mathbf{k}$,

$$\oint_C \mathbf{F}.d\mathbf{r} = \iint_R (\nabla \wedge \mathbf{F}).d\mathbf{S}, \tag{4.4}$$

which shows that Green's theorem in the plane is a particular case of Stokes's theorem (4.18).

Problem 4.1 Show that the area of the region R is given by

$$A = \tfrac{1}{2}\oint (x\,dy - y\,dx).$$

Solution. Putting $P = -y$, $Q = x$ in (4.1) we get

$$\oint_C (x\,dy - y\,dx) = \iint_R \left[\frac{\partial}{\partial x}(x) + \frac{\partial}{\partial y}(y)\right] dx\,dy$$

$$= \iint_R 2\,dx\,dy = 2A.$$

(Alternative forms for A are $\oint_C x\,dy$ and $-\oint_C y\,dx$, as is found by taking $P = 0$, $Q = x$, and $P = -y$, $Q = 0$, respectively.) □

Problem 4.2 Evaluate

$$I = \oint_C [y(2xy-1)dx + x(2xy+1)dy],$$

where C is the circle $x^2 + y^2 = 1$ described in the anticlockwise sense (i) directly, (ii) using Green's theorem in the plane.

Solution. (i) On C we can put $x = \cos t$, $y = \sin t$, $(0 \leqslant t \leqslant 2\pi)$. Then $dx = -\sin t\,dt$, $dy = \cos t\,dt$, and I reduces to

$$I = \int_0^{2\pi} (-2\sin^3 t\cos t + 2\cos^3 t\sin t + 1)\,dt$$

$$= \left| -\tfrac{1}{2}\sin^4 t - \tfrac{1}{2}\cos^4 t + t \right|_0^{2\pi} = 2\pi.$$

(ii) If R denotes the disc $x^2 + y^2 \leqslant 1$, by (4.1)

$$I = \iint_R \left\{\frac{\partial}{\partial x}[x(2xy+1)] - \frac{\partial}{\partial y}[y(2xy-1)]\right\} dx\,dy$$

$$= \iint_R [(4xy+1) - (4xy-1)]\,dx\,dy$$

$$= \iint_R 2\,dx\,dy = 2\pi,$$

since π is the area of R. □

Problem 4.3 If $\mathbf{F} = (x^2+y^2)^{-1}(-y\mathbf{i}+x\mathbf{j})$, evaluate $\oint_C \mathbf{F}.d\mathbf{r}$ where C is any simple closed curve about the origin described in the anticlockwise sense, Calculate $\nabla \wedge \mathbf{F}$. Why is (4.1) apparently violated?

Solution. Since

$$\frac{\partial}{\partial x}\left(\tan^{-1}\frac{y}{x}\right) = \frac{-y}{x^2+y^2}, \quad \frac{\partial}{\partial y}\left(\tan^{-1}\frac{y}{x}\right) = \frac{x}{x^2+y^2}$$

we have $\mathbf{F} = \nabla\tan^{-1}(y/x) = \nabla\theta$, where θ is the polar angle of the point (x, y) relative to the x-axis. Hence (*cf.* Problem 2.7)

$$\oint_C \mathbf{F}.d\mathbf{r} = \oint_C \nabla\theta.d\mathbf{r} = \oint_C d\theta = 2\pi.$$

(Note that although $\mathbf{F}$ is expressible as the gradient of a scalar, θ, the line integral does not vanish. This is because θ is not a *single-valued* function of x and y.)

We find, for $(x, y) \neq (0, 0)$,

$$\nabla \wedge \mathbf{F} = \begin{vmatrix} \mathbf{i} & \mathbf{j} & \mathbf{k} \\ \partial/\partial x & \partial/\partial y & \partial/\partial z \\ \dfrac{-y}{x^2+y^2} & \dfrac{x}{x^2+y^2} & 0 \end{vmatrix} = 0, \tag{4.5}$$

on expanding and calculating the derivatives.

Denoting by R the region enclosed by C, it is seen that the conditions of Green's theorem in the plane are not satisfied by the vector $\mathbf{F} = P\mathbf{i}+Q\mathbf{j}$ at all points of R, since P, Q are not continuous (or even defined) at $(0, 0)$. Hence (4.1) and (4.4) do not apply.

Let C_1 be a second simple closed curve about O, entirely contained in R, and let R' be the region bounded by C_1 and C (Fig. 4.1).

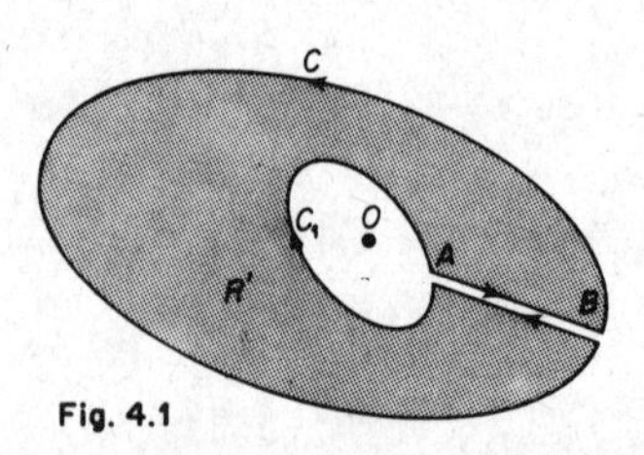

Fig. 4.1

Introduce a 'cross-cut' AB from C_1 to C The closed curve formed by C, BA, C_1, AB, as indicated by arrows, may be taken as a contour for Green's theorem with R' as 'enclosed' region. Since the contour integrals along BA and AB cancel, we have according to (4.4),

$$\left(\oint_C + \oint_{C_1}\right)\mathbf{F}\,.\,d\mathbf{r} = \iint_{R'} (\nabla \wedge \mathbf{F})\,.\,\mathbf{k}\; dx\, dy = 0. \tag{4.6}$$

This represents a correct application of the theorem; by reversing the direction of C_1 we find that the tangential line integrals of $\mathbf{F}$ around the two curves C, C_1, described in the *same* sense, are equal. We cannot conclude from (4.6) that either line integral vanishes, despite (4.4), (4.5).

□

4.2 Gauss's Divergence Theorem Let a three-dimensional region V be bounded by a closed surface S. If a vector function $\mathbf{F}$ is continuous and has continuous partial derivatives in V and on S, then the outward flux of $\mathbf{F}$ through S is equal to the volume integral of div $\mathbf{F}$ taken over V. That is,

$$\int_S \mathbf{F}\,.\,d\mathbf{S} = \int_V \operatorname{div} \mathbf{F}\; dV, \tag{4.7}$$

where $d\mathbf{S}$ has the direction of the outward normal. This result is known as Gauss's divergence theorem, and is a three-dimensional analogue of Green's theorem in the plane.

We give a sketch proof for the case where every line which crosses V and is parallel to a coordinate axis, meets S in just two points. Suppose that the line y = constant, z = constant meets S in the points (x_1, y, z), (x_2, y, z), $(x_1 < x_2)$. Then if $\mathbf{F} = F_x\,\mathbf{i} + F_y\,\mathbf{j} + F_z\,\mathbf{k}$, the integrand on the right in (4.7) is $(\partial F_x/\partial x) + (\partial F_y/\partial y) + (\partial F_z/\partial z)$, and for the contribution from the first term we have

$$\int_V \frac{\partial F_x}{\partial x}\, dV = \iint_{S_0} \left(\int_{x_1}^{x_2} \frac{\partial F_x}{\partial x}\, dx\right) dy\, dz = \iint_{S_0} [F_x(x_2, y, z) - F_x(x_1, y, z)]\; dy\, dz$$

where S_0 is the projection of S on the plane $x = 0$. The elements $d\mathbf{S}_1$, $d\mathbf{S}_2$ of S located respectively at (x_1, y, z), (x_2, y, z) and having rectangular projections $dy\, dz$ on $x = 0$ (Fig. 4.2), are such that $d\mathbf{S}_1\,.\,\mathbf{i} = -dy\, dz$, $d\mathbf{S}_2\,.\,\mathbf{i} = dy\, dz$. Hence the third integral in (4.8) is

$$\int_{S_2} F_x \mathbf{i}\,.\,d\mathbf{S}_2 + \int_{S_1} F_x \mathbf{i}\,.\,d\mathbf{S}_1 = \int_S F_x \mathbf{i}\,.\,d\mathbf{S},$$

where S_1 and S_2 are the parts of S containing all points (x_1, y, z) and (x_2, y, z), respectively.

The other terms in the right-hand integral (4.7) are dealt with by projection onto the planes $y = 0$ and $z = 0$, and on adding the results so obtained we find, as required,

$$\int_V \operatorname{div} \mathbf{F}\; dV = \int_S (F_x \mathbf{i} + F_y \mathbf{j} + F_z \mathbf{k})\,.\,d\mathbf{S} = \int_S \mathbf{F}\,.\,d\mathbf{S}.$$

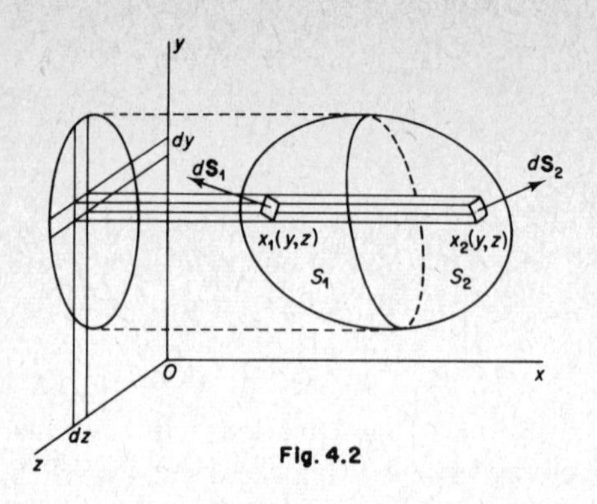

Fig. 4.2

Equation (4.7) holds in much more general cases than are covered in this proof. For example, V may be bounded by a number of separate closed surfaces $S^0, S^1, S^2, \ldots, S^n$, of which $S^1, S^2, \ldots, S^n$ are exterior to one another but are interior to S^0. (The case $n = 2$ is illustrated in Fig. 4.3.) In applying Gauss's theorem in such cases, the surface integral is taken to mean the sum of $n+1$ surface integrals over the separate parts of the boundary of V, with $d\mathbf{S}$ always directed away from V.

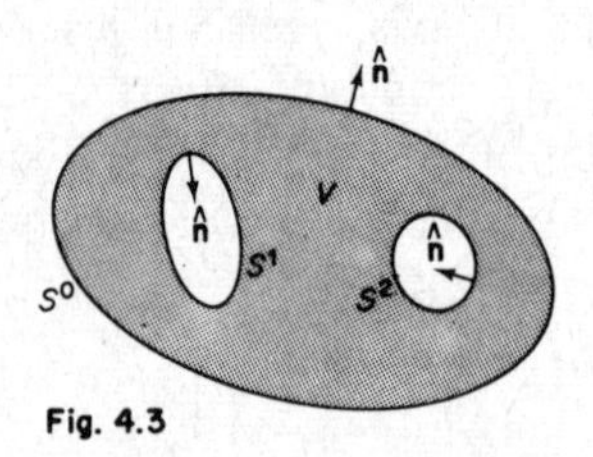

Fig. 4.3

Problem 4.4 If $\mathbf{F} = xz\mathbf{i}+3xy\mathbf{j}-2z\mathbf{k}$, evaluate $\int_S \mathbf{F}.d\mathbf{S}$ by means of Gauss's theorem when (i) S is the closed cylinder bounded by the surface $x^2+y^2 = 1$ and the planes $z = 0$, $z = 3$; (ii) S is the curved cylindrical surface $x^2+y^2 = 1, 0 < z < 3$.

Solution. (i) Let V denote the solid cylinder $x^2+y^2 \leqslant 1$, $0 \leqslant z \leqslant 3$. Then

$$\int_S \mathbf{F}.d\mathbf{S} = \int_V \left[\frac{\partial}{\partial x}(xz)+\frac{\partial}{\partial y}(3xy)+\frac{\partial}{\partial z}(-2z)\right] dV = \int_V (z+3x-2)\, dV.$$

Transforming to cylindrical polar coordinates: $x = R\cos\psi$, $y = R\sin\psi$, $z = z$, since $dV = R\,dR\,d\psi\,dz$ (Fig. 5.1),

$$\int_S \mathbf{F}.d\mathbf{S} = \int_0^3\int_0^{2\pi}\int_0^1 (z+3R\cos\psi-2)R\,dR\,d\psi\,dz$$

$$= \int_0^3 \int_0^{2\pi} \left| \tfrac{1}{2}(z-2)R^2 + R^3 \cos\psi \right|_0^1 d\psi\, dz$$

$$= \int_0^3 \int_0^{2\pi} (\tfrac{1}{2}z - 1 + \cos\psi)\, d\psi\, dz$$

$$= 2\pi \int_0^3 (\tfrac{1}{2}z - 1)\, dz = -\tfrac{3}{2}\pi.$$

(ii) Let S_1, S_2 denote the plane ends of the closed cylinder in (i), i.e. $x^2+y^2 < 1$, $z = 0$, and $x^2+y^2 < 1$, $z = 3$, respectively. On S_1, the normal directed away from V is $-\mathbf{k}$, while on S_2 it is $\mathbf{k}$. Hence,

$$\int_{S_1} \mathbf{F}.d\mathbf{S} = \int_{S_1} \mathbf{F}.(-\mathbf{k})\, dS = \int_{S_1} 2z\, dS = 0, \tag{4.9}$$

since $z = 0$ on S_1. Also, on S_2, $\mathbf{F}.\mathbf{k} = -2z = -6$,

$$\int_{S_2} \mathbf{F}.d\mathbf{S} = \int_{S_2} \mathbf{F}.\mathbf{k}\, dS = -6 \int_{S_2} dS = -6\pi, \tag{4.10}$$

since the area of S_2 is π. But by (i)

$$\left(\int_S + \int_{S_1} + \int_{S_2}\right) \mathbf{F}.d\mathbf{S} = -\tfrac{3}{2}\pi,$$

and so by (4.9), (4.10),

$$\int_S \mathbf{F}.d\mathbf{S} = -\tfrac{3}{2}\pi - (-6\pi) = \tfrac{9}{2}\pi. \qquad \square$$

Problem 4.5 Use Gauss's theorem to evaluate $\int_S \mathbf{F}.d\mathbf{S}$, where $\mathbf{F} = xy\mathbf{i} + \mathbf{j} + (z+1)\mathbf{k}$ and S is the curved surface of the hemisphere $x^2+y^2+z^2 = 9$, $z \geqslant 0$.

Solution. Let S_0 denote the disc $x^2+y^2 < 9$, $z = 0$, and V the solid hemisphere bounded by S and S_0. Then by (4.7),

$$\int_S \mathbf{F}.d\mathbf{S} + \int_{S_0} \mathbf{F}.d\mathbf{S}_0 = \int_V \nabla.\mathbf{F}\, dV = \int_V (y+1)\, dV. \tag{4.11}$$

Writing $d\mathbf{S}_0 = -\mathbf{k}\, dx\, dy$, we have since $\mathbf{F}.\mathbf{k} = z+1 = 1$ on S_0,

$$\int_{S_0} \mathbf{F}.d\mathbf{S}_0 = -\int_{S_0} \mathbf{F}.\mathbf{k}\, dx\, dy = -9\pi, \tag{4.12}$$

the area of S_0 being 9π.

For the volume integral in (4.11), we may transform to spherical polar coordinates (2.36). The volume element is $dV = r^2 \sin\theta\, dr\, d\theta\, d\psi$ (see Problem 2.22). Hence

$$\int_V \nabla.\mathbf{F}\, dV = \int_V (y+1)\, dV = \int_0^{2\pi} \int_0^{\frac{1}{2}\pi} \int_0^3 (r \sin\theta \sin\psi + 1) r^2 \sin\theta\, dr\, d\theta\, d\psi.$$

Carrying out the ψ integration first shows that the contribution from the first bracketed term is zero, and that from the second bracketed term is 18π, the volume of V. Therefore, by (4.11), (4.12),

$$\int_S \mathbf{F}.d\mathbf{S} = 18\pi - (-9\pi) = 27\pi,$$

if $d\mathbf{S}$ points away from the origin. If the opposite sense for $d\mathbf{S}$ is taken, the corresponding value is -27π. □

Problem 4.6 Verify Gauss's theorem for the vector $\mathbf{F} = f(r)\hat{\mathbf{r}}$, where f possesses a continuous derivative, over the region V which is bounded by the spheres $r = a$, $r = b$, $(a < b)$.

Solution. Denoting the inner and outer bounding spheres by S_1 and S_2 respectively, we have that the outward unit normal vector is $-\hat{\mathbf{r}}$ at a point $\mathbf{r}$ on S_1, and $+\hat{\mathbf{r}}$ at a point $\mathbf{r}$ on S_2. Hence,

$$\int_S \mathbf{F}.d\mathbf{S} = \Big(\int_{S_1} + \int_{S_2}\Big)\mathbf{F}.d\mathbf{S} = \int_{S_1} \mathbf{F}.(-\hat{\mathbf{r}})\,dS + \int_{S_2} \mathbf{F}.\hat{\mathbf{r}}\,dS$$

$$= -\int_{S_1} f(a)\,dS + \int_{S_2} f(b)\,dS = -f(a)(4\pi a^2) + f(b)(4\pi b^2)$$

$$= 4\pi[b^2 f(b) - a^2 f(a)]. \tag{4.13}$$

By Problem 3.4,

$$\text{div}\,\mathbf{F} = \text{div}[r^{-1}f(r)\mathbf{r}] = r\frac{d}{dr}(r^{-1}f) + 3(r^{-1}f)$$

$$= \frac{df}{dr} + \frac{2}{r}f.$$

Writing dV in spherical polar coordinates, as in the last problem, we get

$$\int_V \text{div}\,\mathbf{F}\,dV = \int_0^{2\pi}\int_0^{\pi}\int_a^b \left(\frac{df}{dr} + \frac{2}{r}f\right) r^2 \sin\theta\,dr\,d\theta\,d\psi$$

$$= \int_0^{2\pi}\int_0^{\pi}\left[\int_a^b \frac{d}{dr}(r^2 f)\,dr\right]\sin\theta\,d\theta\,d\psi$$

$$= 4\pi\,\Big|\,r^2 f(r)\,\Big|_a^b = 4\pi[b^2 f(b) - a^2 f(a)]. \tag{4.14}$$

Comparing (4.13), (4.14), we find as required

$$\int_S \mathbf{F}.d\mathbf{S} = \int_V \text{div}\,\mathbf{F}\,dV.$$ □

Problem 4.7 Show that if $\mathbf{B} = \nabla \wedge \mathbf{A}$, where $\mathbf{A}$ is twice continuously differentiable, the flux of **B** out through an arbitrary closed surface S is zero.

Solution. The flux of **B** through S is

$$\int_S \mathbf{B}.d\mathbf{S} = \int_V \nabla.\mathbf{B}\, dV = \int_V \nabla.(\nabla\wedge\mathbf{A})\, dV = 0$$

(since $\nabla.(\nabla\wedge\mathbf{A}) \equiv 0$ (Problem 3.15)). □

Problem 4.8 Show that if ϕ is continuously differentiable in the region V and on its boundary S, then

$$\int_S \phi\, d\mathbf{S} = \int_V \nabla\phi\, dV. \tag{4.15}$$

Evaluate $\int_S d\mathbf{S}$.

Solution. Let $\hat{\mathbf{a}}$ be an arbitrary constant unit vector. The component of the left-hand integral in (4.15), in the direction $\hat{\mathbf{a}}$, is

$$\hat{\mathbf{a}}.\int_S \phi\, d\mathbf{S} = \int_S \phi\hat{\mathbf{a}}.d\mathbf{S} = \int_V \nabla.(\phi\hat{\mathbf{a}})\, dV$$

$$= \int_V [\phi\nabla.\hat{\mathbf{a}}+\hat{\mathbf{a}}.\nabla\phi]\, dV,$$

by identity (ii), § 3.3. Since $\hat{\mathbf{a}}$ is constant, this becomes

$$\int_V \hat{\mathbf{a}}.\nabla\phi\, dV = \hat{\mathbf{a}}.\int_V \nabla\phi\, dV.$$

Hence both sides of (4.15) have the same component in the direction of $\hat{\mathbf{a}}$, which is arbitrary, and the result follows immediately.

(*Note*: the validity of moving $\hat{\mathbf{a}}$ inside or outside an integral sign in the above manner is formally shown by first expressing the integral as the appropriate limit of a sum.)

Taking $\phi = 1$ in (4.15) we get

$$\int_S d\mathbf{S} = 0,$$

S being an arbitrary closed surface. □

Problem 4.9 Show that if **F** is continuously differentiable,

$$\int_S \mathbf{F}\wedge d\mathbf{S} = -\int_V (\nabla\wedge\mathbf{F})\, dV. \tag{4.16}$$

Solution. The component of the left-hand integral in the direction of an arbitrary constant unit vector $\hat{\mathbf{a}}$ is

$$\hat{\mathbf{a}}.\int_S \mathbf{F}\wedge d\mathbf{S} = \int_S \hat{\mathbf{a}}.(\mathbf{F}\wedge d\mathbf{S}) = \int_S (\hat{\mathbf{a}}\wedge\mathbf{F}).d\mathbf{S}$$

(interchanging dot and cross). By Gauss's theorem this becomes

$$\hat{\mathbf{a}}.\int_S \mathbf{F}\wedge d\mathbf{S} = \int_V \nabla.(\hat{\mathbf{a}}\wedge\mathbf{F})\, dV$$

$$= \int_V [\mathbf{F}.(\nabla \wedge \hat{\mathbf{a}}) - \hat{\mathbf{a}}.(\nabla \wedge \mathbf{F})]\, dV = -\hat{\mathbf{a}}.\int_V (\nabla \wedge \mathbf{F})\, dV,$$

where we have used § 3.3 (iv) and the fact that $\hat{\mathbf{a}}$ is constant. Hence the components of each side of (4.16) are equal in the arbitrary direction $\hat{\mathbf{a}}$, which proves the result. □

Problem 4.10 (i) Use (4.15) to obtain a surface integral formula for $\nabla\phi$. (ii) Find corresponding formulae for $\nabla . \mathbf{F}$ and $\nabla \wedge \mathbf{F}$.

Solution. Let V be a small volume containing an arbitrary point P. Assuming $\nabla\phi$ to be continuous, we have

$$\int_V \nabla\phi\, dV = \int_V [(\nabla\phi)_P + \epsilon]\, dV = (\nabla\phi)_P V + \int_V \epsilon\, dV,$$

where the suffix P denotes that the value at P is to be taken, and ϵ is a vector whose magnitude is as small as we please throughout V, provided the dimensions of V are sufficiently small. Dividing by V and using (4.15),

$$\frac{1}{V}\int_S \phi\, d\mathbf{S} = (\nabla\phi)_P + \frac{1}{V}\int_V \epsilon\, dV.$$

The second term on the right tends to zero, as $V \to 0$, because

$$\left|\frac{1}{V}\int_V \epsilon\, dV\right| \leqslant \frac{1}{V}|\epsilon|_{\max} V = |\epsilon|_{\max},$$

where max denotes the maximum value in V. Hence at any point P,

$$\nabla\phi = \lim_{V \to 0} \frac{1}{V}\int_S \phi\, d\mathbf{S},$$

where S is any closed surface enclosing P and V is the interior volume.

(ii) By Gauss's theorem, if $\nabla . \mathbf{F}$ is continuous,

$$\frac{1}{V}\int_S \mathbf{F}.d\mathbf{S} = \frac{1}{V}\int_V \nabla . \mathbf{F}\, dV,$$

and using a similar procedure to that in (i) we get at any point P,

$$\nabla . \mathbf{F} = \lim_{V \to 0} \frac{1}{V}\int_S \mathbf{F}.d\mathbf{S},$$

S and V having the same meanings as before. This shows that the divergence of a vector is the 'outward flux per unit volume' at each field point. Finally, from (4.16) we obtain in the same way

$$\nabla \wedge \mathbf{F} = -\lim_{V \to 0} \frac{1}{V}\int_S \mathbf{F} \wedge d\mathbf{S}.$$ □

Problem 4.11 Heat conduction in a uniform body takes place in the opposite direction to the gradient of the temperature $\phi(x, y, z, t)$, the heat energy crossing any surface S per unit time being $-k\int_S \nabla\phi \,.d\mathbf{S}$, where k ($k > 0$) is the *thermal conductivity* constant. (i) Show that if a steady state has been attained (i.e. the temperature is independent of time), then $\nabla^2\phi = 0$ in any region free of heat sources. (ii) If the body is bounded by the spheres $x^2+y^2+z^2 = 1$, $x^2+y^2+z^2 = 4$, which are maintained at constant temperatures ϕ_1, ϕ_2, respectively, find the steady state temperature at any interior point.

Solution. By conservation of energy, the rate of increase of heat energy in a region V enclosed by any closed surface S is equal to the rate at which heat flows inwards across the boundary, provided that there are no heat sources (chemical, electrical, etc.) in V.

The heat energy in an element ΔV of V is $c\rho\phi\,\Delta V$, where ρ is the density and c the specific heat. Therefore,

$$\text{rate of increase of heat in } \Delta V = c\rho\frac{\partial\phi}{\partial t}\Delta V,$$

and by summing over all elements and taking the limit as all $\Delta V \to 0$ we get

$$\text{rate of increase of heat in } V = c\rho\int_V (\partial\phi/\partial t)dV,$$

The *inward* flow across the boundary is, by Gauss's theorem,

$$k\int_S \nabla\phi \,.\, d\mathbf{S} = k\int_V \nabla\,.\,(\nabla\phi)\, dV = k\int_V \nabla^2\phi\, dV.$$

Hence, putting $\kappa = k/\rho c$ (the *diffusivity*), by conservation we must have

$$\int_V [\kappa\nabla^2\phi-(\partial\phi/\partial t)]\, dV = 0.$$

Assuming the integrand is continuous, it must vanish identically, since V is arbitrary. (Otherwise, taking V to be a neighbourhood in which the integrand is everywhere positive, or everywhere negative, leads to a contradiction.) Thus,

$$\kappa\nabla^2\phi = \partial\phi/\partial t,$$

which is the equation of heat conduction for a uniform body. In a steady state, $\partial\phi/\partial t = 0$, and so $\nabla^2\phi = 0$.

(ii) By symmetry, ϕ is a function of $r = (x^2+y^2+z^2)^{\frac{1}{2}}$ only, $\phi = f(r)$, say. Using the result in Problem 3.13(i),

$$\frac{d^2f}{dr^2}+\frac{2}{r}f = 0.$$

Multiplying by the integrating factor r^2,

$$\frac{d}{dr}\left(r^2\frac{df}{dr}\right) = 0, \qquad r^2\frac{df}{dr} = C_1,$$

where C_1 is a constant of integration. A further integration gives

$$\phi = f(r) = -\frac{C_1}{r}+C_2, \qquad (C_2 = \text{constant}). \tag{4.17}$$

Putting $\phi = \phi_1$ when $r = 1$, and $\phi = \phi_2$ when $r = 2$, gives

$$\phi_1 = -C_1+C_2, \qquad \phi_2 = -\tfrac{1}{2}C_1+C_2.$$

Solving for C_1 and C_2, and substituting in (4.17) we get in $1 \leqslant r \leqslant 2$,

$$\phi = \frac{2(\phi_1-\phi_2)}{r}+2\phi_2-\phi_1. \qquad \square$$

4.3 Stokes's Theorem We say that a surface is *open* if it is bounded by a closed curve. In this sense a hemisphere is open but a sphere is not. Let a unit normal $\hat{\mathbf{n}}$ be constructed at any point P on a surface S, and let P move in a closed path on S. Unless $\hat{\mathbf{n}}$ changes sign discontinuously during the circuit we find, for most familiar types of surface, that the initial and final directions of $\hat{\mathbf{n}}$ are always the same. Such surfaces (unlike the Möbius strip, for example) are *two-sided* or *orientable*. We shall consider here only two-sided surfaces.

A surface S is *simply-connected* if every closed curve drawn on S can be shrunk continuously to a point without leaving S. Thus, in the case of plane surfaces, the disc is simply-connected but the annulus (disc with a hole) is not.

Let $\hat{\mathbf{n}}$ be the normal on a particular side of an open (simply-connected) surface S. A small loop about any point P_0 on S is described *positively* with respect to $\hat{\mathbf{n}}$ if the sense is that of the rotation needed to drive a right-handed screw forward along $\hat{\mathbf{n}}$ at P. By allowing the loop to expand until it coincides with the boundary C of S, we obtain a natural definition of the positive sense of description of C with respect to $\hat{\mathbf{n}}$.

Stokes's Theorem Let S be an open (simply-connected) surface whose boundary is the closed curve C, and let $\mathbf{F} = F_x\mathbf{i}+F_y\mathbf{j}+F_z\mathbf{k}$ be continuous and have continuous partial derivatives in S and on C. Then

$$\oint_C \mathbf{F}.d\mathbf{r} = \int_S (\nabla\wedge\mathbf{F}).d\mathbf{S}, \tag{4.18}$$

where C is described positively with respect to $d\mathbf{S}$.

The full proof of this theorem is rather long, and may be found in standard texts on vector analysis*. The case where S is a region of the xy-plane is identical to Green's theorem in the plane (4.1).

Problem 4.12 If $\mathbf{F} = y\mathbf{i} - x\mathbf{j} + yz\mathbf{k}$, evaluate $\int_S \text{curl } \mathbf{F} \cdot d\mathbf{S}$ where S is the part of the surface $z = 2(x^2+y^2)$ for which $z \leqslant \frac{1}{2}$: (i) by direct integration, (ii) by means of Stokes's theorem.

Solution. The surface in question is a paraboloid of revolution about Oz (Fig. 4.4), and the integral is to be taken over that part whose projection

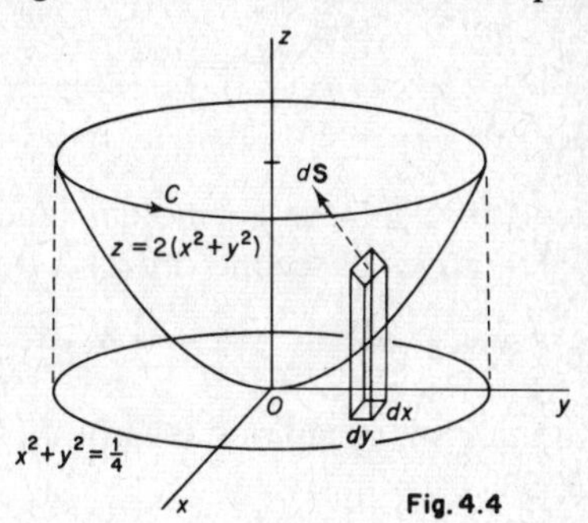

Fig. 4.4

on the xy-plane is the circular region $x^2+y^2 \leqslant \frac{1}{4}$. For definiteness, we evaluate the integral taking $d\mathbf{S}$ to point upwards (have a positive z component); the positive direction of the boundary is shown in the diagram.

(i)

$$\text{curl } \mathbf{F} = \begin{vmatrix} \mathbf{i} & \mathbf{j} & \mathbf{k} \\ \partial/\partial x & \partial/\partial y & \partial/\partial z \\ y & -x & yz \end{vmatrix} = z\mathbf{i} - 2\mathbf{k},$$

and a normal to S is

$$\mathbf{n} = \nabla[z - 2(x^2+y^2)] = -4x\mathbf{i} - 4y\mathbf{j} + \mathbf{k}.$$

By the method of Problem 2.16,

$$\int_S \text{curl } \mathbf{F} \cdot d\mathbf{S} = \iint_{x^2+y^2 \leqslant \frac{1}{4}} \frac{\text{curl } \mathbf{F} \cdot \mathbf{n}}{\mathbf{k} \cdot \mathbf{n}} \, dx \, dy$$

$$= -\iint_{x^2+y^2 \leqslant \frac{1}{4}} (4xz+2) \, dx \, dy = -\tfrac{1}{2}\pi,$$

as we find on substituting for z in terms of x and y on S, and noting that the

* See, for example, *Vector Analysis* by L. Marder (George Allen and Unwin).

contribution from the first term in the integrand must vanish since the term is an odd function of x.

(ii) The boundary curve C may be expressed parametrically as $x = \frac{1}{2}\cos\psi$, $y = \frac{1}{2}\sin\psi$, with ψ increasing from 0 to 2π. Hence by Stokes's theorem,

$$\int_S \text{curl}\,\mathbf{F}\,.\,d\mathbf{S} = \int_C \mathbf{F}\,.\,d\mathbf{r} = \int_C (y\,dx - x\,dy)$$

$$= \tfrac{1}{4}\int_0^{2\pi} \sin\psi(-\sin\psi\,d\psi) - \cos\psi(\cos\psi\,d\psi)$$

$$= -\tfrac{1}{4}\int_0^{2\pi} d\psi = -\tfrac{1}{2}\pi,$$

in agreement with (i). □

Problem 4.13 Prove that if **A** is solenoidal (and is continuous with continuous partial derivatives of second order),

$$\int_S \nabla^2\mathbf{A}\,.\,d\mathbf{S} = -\oint_C \text{curl}\,\mathbf{A}\,.\,d\mathbf{r},$$

where S and C have the same meanings as in Stokes's theorem.

Solution. By § 3.3(vii),

$$\nabla^2\mathbf{A} = \text{grad div}\,\mathbf{A} - \text{curl curl}\,\mathbf{A}$$

$$= -\text{curl curl}\,\mathbf{A},$$

when **A** is solenoidal (i.e. div **A** = 0). Hence by Stokes's theorem,

$$\int_S \nabla^2\mathbf{A}\,.\,d\mathbf{S} = -\int_S \text{curl curl}\,\mathbf{A}\,.\,d\mathbf{S}$$

$$= -\oint_C \text{curl}\,\mathbf{A}\,.\,d\mathbf{r}.$$ □

Problem 4.14 Show that for any continuously differentiable scalar ϕ,

$$\oint_C \phi\,d\mathbf{r} = -\int_S \nabla\phi \wedge d\mathbf{S}, \tag{4.19}$$

in the notation of Stokes's theorem.

Solution. Form the component of the left-hand integral in the direction of an arbitrary constant unit vector $\hat{\mathbf{a}}$. Thus,

$$\hat{\mathbf{a}}\,.\oint_C \phi\,d\mathbf{r} = \oint_C \phi\hat{\mathbf{a}}\,.\,d\mathbf{r} = \int_S [\nabla \wedge (\phi\hat{\mathbf{a}})]\,.\,d\mathbf{S}$$

$$= \int_S [\phi\nabla \wedge \hat{\mathbf{a}} + \nabla\phi \wedge \hat{\mathbf{a}}]\,.\,d\mathbf{S} = -\int_S \hat{\mathbf{a}}\,.(\nabla\phi \wedge d\mathbf{S}),$$

by a standard property of scalar triple products, since $\nabla \wedge \hat{\mathbf{a}} = 0$. Hence,

$$\hat{\mathbf{a}}\,.\oint_C \phi\,d\mathbf{r} = -\hat{\mathbf{a}}\,.\int_S (\nabla\phi \wedge d\mathbf{S}),$$

and the result follows because the direction of $\hat{\mathbf{a}}$ is arbitrary. □

Problem 4.15 Prove that

$$\int_S \mathbf{r} \wedge d\mathbf{S} = -\tfrac{1}{2} \oint_C r^2 \, d\mathbf{r},$$

where $\mathbf{r} = x\mathbf{i}+y\mathbf{j}+z\mathbf{k}$.

Solution. Setting $\phi = \frac{1}{2}r^2$ in (4.19) the result is immediate, since

$$\nabla(\tfrac{1}{2}r^2) = r\nabla r = r\hat{\mathbf{r}} = \mathbf{r}. \qquad \square$$

Problem 4.16 If $\mathbf{q}(x, y, z, t)$ denotes the velocity at the point (x, y, z) at time t in a moving fluid, the integral $\oint_C \mathbf{q} \,.\, d\mathbf{r}$ is called the *circulation* about the arbitrary closed curve C. (i) Show that for the 'laminar shear flow' given by $\mathbf{q} = \mu y\mathbf{i}$ (μ = constant), the circulation is proportional to the area enclosed by the projection of C on the xy-plane (assuming the projection to be a simple closed curve); (ii) show that if $\mathbf{q} = \kappa R^{-1}\hat{\boldsymbol{\psi}}$ (κ = constant) in cylindrical polar coordinates, where $\hat{\boldsymbol{\psi}}$ is the unit vector in the direction of ψ increasing (with R, z held constant), then the circulation is $2\pi\kappa$ if C comprises a single positive circuit about Oz, and is zero if C does not form a circuit of Oz.

Solution. (i) We find

$$\nabla \wedge \mathbf{q} = \nabla \wedge (\mu y\mathbf{i}) = -\mu\mathbf{k}.$$

By Stokes's theorem, if S is an open surface bounded by C,

$$\oint_C \mathbf{q} \,.\, d\mathbf{r} = -\int_S \mu\mathbf{k} \,.\, d\mathbf{S} = -\mu \int_S \mathbf{k} \,.\, \hat{\mathbf{n}} \, dS$$

where $\hat{\mathbf{n}}$ is the normal for which C is described positively. As in Problem 2.16, we can put $\hat{\mathbf{n}} \,.\, \mathbf{k} \, dS = \pm dx\, dy$, the sign being that of $\hat{\mathbf{n}} \,.\, \mathbf{k}$, and so the circulation is $\pm\mu S_0$, where S_0 is the (area of) the projection of S on the xy-plane.

Note that we have assumed that $\hat{\mathbf{n}} \,.\, \mathbf{k}$ does not change sign over S, which means that we choose a surface which is nowhere 'vertical' (parallel to Oz).

(ii) By a formula for $\nabla \wedge \mathbf{q}$ in cylindrical polar coordinates (5.27) we have

$$\nabla \wedge \mathbf{q} = \frac{1}{R}\begin{vmatrix} \hat{\mathbf{R}} & R\hat{\boldsymbol{\psi}} & \hat{\mathbf{z}} \\ \partial/\partial R & \partial/\partial\psi & \partial/\partial z \\ 0 & \kappa & 0 \end{vmatrix} = 0, \qquad (R \neq 0). \tag{4.20}$$

If C makes a positive circuit of Oz (Fig. 4.4), we cannot apply Stokes's theorem directly, since every surface bounded by C contains a point on Oz, where $\nabla \wedge \mathbf{q}$ is not continuous (or defined). Let C_1 be any second positive circuit of Oz (not intersecting C), and let S be a surface bounded by C and C_1. Introduce any 'cross-cut' AB on S from C_1 to C, so that S becomes simply-connected and has the simple closed curve C, BA, $-C_1$, AB as boundary. By Stokes's theorem,

$$\left(\int_C + \int_{BA} - \int_{C_1} + \int_{AB}\right)\mathbf{q}\,.\,d\mathbf{r} = \int_S (\nabla\wedge\mathbf{q})\,.\,d\mathbf{S} = 0,$$

because of (4.20). The contributions along BA and AB cancel, and so

$$\oint_C \mathbf{q}\,.\,d\mathbf{r} = \int_{C_1} \mathbf{q}\,.\,d\mathbf{r},$$

which shows that the circulation about C is the same as that about any other positive circuit C_1.

If we take C_1 to be a circle $R = a$, $z =$ constant, we get

$$\int_{C_1} \mathbf{q}\,.\,d\mathbf{r} = \int_0^{2\pi} \frac{\kappa}{a}\hat{\boldsymbol{\psi}}\,.\,(a\,d\psi\,\hat{\boldsymbol{\psi}}) = \kappa\int_0^{2\pi} d\psi = 2\pi\kappa,$$

and so this is the circulation about every single positive circuit about Oz.

□

4.4 Green's Integral Identities Let u and v be continuous scalar functions with continuous first and second partial derivatives. If we write $\mathbf{F} = u\nabla v$ in Gauss's theorem we get

$$\int_V \nabla\,.\,(u\nabla v)\,dV = \int_S u\nabla v\,.\,\hat{\mathbf{n}}\,dS.$$

Expanding the integrand on the left and noting that the directional derivative of v along the outward normal direction $\mathbf{n}$ is $\partial v/\partial n = \nabla v\,.\,\hat{\mathbf{n}}$,

$$\int_V [u\nabla^2 v + (\nabla u)\,.\,(\nabla v)]\,dV = \int_S u\frac{\partial v}{\partial n}\,dS, \tag{4.21}$$

which is *Green's first identity*.

Interchanging u and v in (4.21) and subtracting the two results gives *Green's second* (or *reciprocal*) *identity*:

$$\int_V (u\nabla^2 v - v\nabla^2 u)\,dV = \int_S \left(u\frac{\partial v}{\partial n} - v\frac{\partial u}{\partial n}\right)dS. \tag{4.22}$$

In some applications the derivatives on the right are not continuous across the boundary S. In this case, the values on the inner side are to be taken, although $\partial/\partial n$ denotes differentiation in the outward normal direction.

Problem 4.17 Let ϕ_1 and ϕ_2 be two solutions of the equation $\nabla^2\phi = 0$ in the region V. Show that if ϕ_1 and ϕ_2 are identical on the boundary S, then they are identical also throughout V (assuming that the two functions have continuous derivatives of up to second order in V and on S).

Solution. Putting $v = u$ in Green's first identity (4.21) we get

$$\int_V [u\nabla^2 u + (\nabla u)^2]\, dV = \int_S u(\partial u/\partial n)\, dS. \tag{4.23}$$

We shall apply this formula to the function $u = \phi_1 - \phi_2$. It is given that:

$$\text{in } V, \quad \nabla^2 u = \nabla^2\phi_1 - \nabla^2\phi_2 = 0;$$

$$\text{on } S, \quad u = \phi_1 - \phi_2 = 0,$$

and so (4.23) reduces to

$$\int_V (\nabla u)^2\, dV = 0. \tag{4.24}$$

But $(\nabla u)^2 \geqslant 0$. We shall show that the equality sign applies at all points. Otherwise, let P be a point in V where $(\nabla u)^2 > 0$; by continuity there is then a neighbourhood of P in V in which this holds, and the contribution to the integral (4.24) from the neighbourhood is strictly positive. As there are no negative contributions, (4.24) is contradicted. Hence $\nabla u \equiv 0$ in V (i.e. all partial derivatives of u vanish), and so u is a constant. But the value of u on the boundary is zero, and therefore $u = \phi_1 - \phi_2 \equiv 0$ in the whole region, which proves the result.

Note that if the equation $\nabla^2\phi = 0$ is replaced by $\nabla^2\phi = f$ in the statement of the problem, where f is a given function, the result remains valid, since the above solution applies without change. The result is known as a *uniqueness theorem*, showing that there can be no more than one solution of the equation $\nabla^2\phi = f$ in a region if ϕ is to take specified values on the boundary. □

Problem 4.18 Let $F_1(x, y, z)\sin k_1 t$ and $F_2(x, y, z)\sin k_2 t$ be solutions of the *wave equation* for $\phi(x, y, z, t)$:

$$\nabla^2\phi = \frac{1}{c^2}\frac{\partial^2\phi}{\partial t^2}, \tag{4.25}$$

in a region V, where k_1, k_2, c are positive constants with $k_1 \neq k_2$. If each solution satisfies the boundary condition

$$\alpha\phi + \beta\frac{\partial\phi}{\partial n} = 0, \quad \text{on } S \tag{4.26}$$

where α and β are functions which take prescribed values on S and do not both vanish together, show that

$$\int_V F_1 F_2\, dV = 0. \tag{4.27}$$

Solution. Substituting $F_1 \sin k_1 t$ for ϕ in (4.25) and cancelling the factor $\sin k_1 t$ we get

$$\nabla^2 F_1 = -(k_1/c)^2 F_1, \qquad \nabla^2 F_2 = -(k_2/c)^2 F_2, \tag{4.28}$$

the second equation following from a similar procedure. By the reciprocal identity (4.22), with $u = F_1$, $v = F_2$, we obtain

$$\int (F_1 \nabla^2 F_2 - F_2 \nabla^2 F_1)\, dV = \int_V \left[F_1\left(-\frac{k_2^2}{c^2} F_2\right) - F_2\left(-\frac{k_1^2}{c^2} F_1\right)\right] dV$$

$$= \int_S \left(F_1 \frac{\partial F_2}{\partial n} - F_2 \frac{\partial F_1}{\partial n}\right) dS. \tag{4.29}$$

Applying (4.26) to each of the given solutions of (4.25) in turn, we have on S,

$$\alpha F_1 + \beta \frac{\partial F_1}{\partial n} = 0, \qquad \alpha F_2 + \beta \frac{\partial F_2}{\partial n} = 0,$$

so that on eliminating α and β,

$$F_1 \frac{\partial F_2}{\partial n} - F_2 \frac{\partial F_1}{\partial n} = 0.$$

Then by (4.29),

$$(k_1^2 - k_2^2) \int_V F_1 F_2 \, dV = 0.$$

Since k_1 and k_2 are positive and distinct, the factor outside the integral sign cancels and the result follows. (The verbal statement of (4.27) is that F_1 and F_2 are *orthogonal functions over* V.) □

EXERCISES

1. Evaluate $\oint_C [y(y-x)\, dx + x(y+x+1)\, dy]$, where C is the ellipse $x = a\cos t$, $y = b \sin t$ described in the positive (anticlockwise) sense, (i) directly, (ii) by using Green's theorem in the plane.

2. Verify Gauss's theorem for the vector $\mathbf{F} = xz\mathbf{i} + yz\mathbf{j} - 2x^2\mathbf{k}$ over the finite region enclosed by part of the paraboloid of revolution $x = R\cos\psi$, $y = R \sin\psi$, $z = R^2$, $0 \leqslant \psi \leqslant 2\pi$, and the plane $z = 1$.

3. Show that if the region V is enclosed by a level surface of $u(x, y, z)$, then for arbitrary $\mathbf{F}$,

$$\int_V (\text{curl}\, \mathbf{F}) \,.\, \text{grad}\, u \, dV = 0.$$

(Assume $\mathbf{F}$ and u to possess continuous partial derivatives of first and second orders, respectively. Consider $\int_S (\mathbf{F} \wedge \operatorname{grad} u) . d\mathbf{S}$, where S is the boundary of V.)

4. Prove that

$$\int_S \hat{\mathbf{r}} \wedge d\mathbf{S} = -\oint_C r \, d\mathbf{r},$$

where $\mathbf{r} = x\mathbf{i} + y\mathbf{j} + z\mathbf{k}$, and S is an open surface bounded by the closed curve C, which is described positively with respect to $d\mathbf{S}$.

5. Evaluate $\int_S \operatorname{curl} \mathbf{F} . d\mathbf{S}$, where S is the open hemispherical surface $x^2 + y^2 + z^2 = a^2$, $z \geqslant 0$, $\mathbf{F} = (1 - ay)\mathbf{i} + 2y^2\mathbf{j} + (x^2 + 1)\mathbf{k}$, and $d\mathbf{S}$ points away from the origin; (i) directly, (ii) by means of Gauss's theorem, (iii) by means of Stokes's theorem.

6. Let ϕ_1 and ϕ_2 each satisfy the equation $\nabla . (\varepsilon \nabla \phi) = 0$ everywhere in the region V enclosed by a closed surface S, where $\varepsilon (> 0)$ is a given function. Show that if $\phi_1 \equiv \phi_2$ on S, then $\phi_1 \equiv \phi_2$ in V, (assuming ε to possess continuous first partial derivatives, and ϕ_1, ϕ_2 continuous second partial derivatives).

Chapter 5

Orthogonal Curvilinear Coordinates

5.1 Curvilinear Coordinates If f, g, h are continuously differentiable functions of x, y, z, and if the equations

$$x = f(u_1, u_2, u_3), \quad y = g(u_1, u_2, u_3), \quad z = h(u_1, u_2, u_3) \qquad (5.1)$$

can be solved uniquely, in some region of space, then to each point (x, y, z) there corresponds a definite set of values (u_1, u_2, u_3), and conversely. The triads of numbers (u_1, u_2, u_3) may then be used as a means of labelling points in place of the cartesian coordinates (x, y, z). In general, the lines along which only one of the u's varies (*coordinate lines*) are curved, and so we call the new system of labelling a *curvilinear coordinate system*. If the coordinate lines always cut perpendicularly, the system is said to be *orthogonal*.

More succinctly, we write (5.1) as $x = x(u_1, u_2, u_3)$, etc., and the inverse relations as

$$u_1 = u_1(x, y, z), \quad u_2 = u_2(x, y, z), \quad u_3 = u_3(x, y, z). \qquad (5.2)$$

Put $\mathbf{r} = x\mathbf{i}+y\mathbf{j}+z\mathbf{k}$, and take differentials

$$d\mathbf{r} = \frac{\partial \mathbf{r}}{\partial u_1} du_1 + \frac{\partial \mathbf{r}}{\partial u_2} du_2 + \frac{\partial \mathbf{r}}{\partial u_3} du_3 . \qquad (5.3)$$

When $du_2 = du_3 = 0$ and $du_1 > 0$, (5.3) represents a differential displacement along the forward tangent to the u_1 coordinate line. Thus, if $\mathbf{e}_1$ is the unit vector in this direction and we write $h_1 = |\partial \mathbf{r}/\partial u_1|$, we have in this case,

$$d\mathbf{r} = \frac{\partial \mathbf{r}}{\partial u_1} du_1 = h_1 \mathbf{e}_1 du_1 . \qquad (5.4)$$

Corresponding results apply for displacements along the tangents to the other two coordinate lines through the point (x, y, z), and (5.3) becomes, for general values of du_1, du_2, du_3,

$$d\mathbf{r} = h_1\, du_1\, \mathbf{e}_1 + h_2\, du_2\, \mathbf{e}_2 + h_3\, du_3\, \mathbf{e}_3 , \qquad (5.5)$$

where $\mathbf{e}_i$ is the forward unit tangent vector to the u_i coordinate line and $h_i = |\partial \mathbf{r}/\partial u_i|$ $(i = 1, 2, 3)$. The h's are functions of the u's, and are called *scale factors*. By taking magnitudes in (5.4) and two similar equations it is seen that h_i is the ratio of the magnitude of displacement, $|d\mathbf{r}|$, to the coordinate increment, du_i, when only the one coordinate u_i is changed.

Denoting by ds the magnitude on either side of (5.5), we have in an orthogonal system:

$$ds^2 = d\mathbf{r} \cdot d\mathbf{r} = h_1^2 du_1^2 + h_2^2 du_2^2 + h_3^2 du_3^2, \tag{5.6}$$

which is called the square of the *line element*. Equation (5.6) is also a consequence of Pythagoras's theorem.

In the remainder of this chapter we shall consider only orthogonal systems. A rectangular *area element* on any of the three coordinate surfaces is easily constructed at any point. For example, for the surface u_1 = constant we take the element whose edges are formed by the vectors $h_2 du_2 \mathbf{e}_2$ and $h_3 du_2 \mathbf{e}_3$; this has area

$$dS_1 = h_2 h_3 du_2 du_3. \tag{5.7}$$

The *volume element* with edges parallel to $\mathbf{e}_1$, $\mathbf{e}_2$, $\mathbf{e}_3$ has volume

$$dV = h_1 h_2 h_3 du_1 du_2 du_3. \tag{5.8}$$

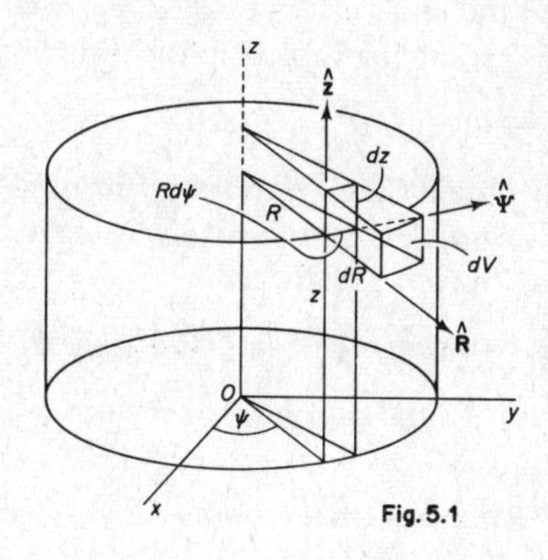

Fig. 5.1

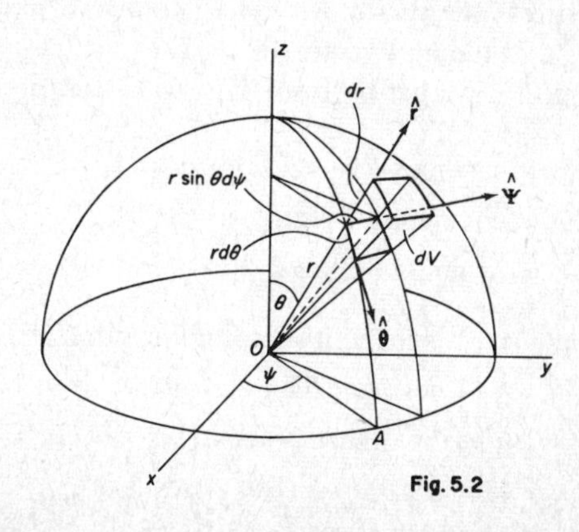

Fig. 5.2

(i) *Cylindrical polar coordinates* (Fig. 5.1):

$$u_1 = R, \quad u_2 = \psi, \quad u_3 = z,$$

$$\mathbf{r} = R\cos\psi\,\mathbf{i} + R\sin\psi\,\mathbf{j} + z\mathbf{k},$$

$$h_1 = |\partial\mathbf{r}/\partial R| = 1, \quad h_2 = |\partial\mathbf{r}/\partial\psi| = R, \quad h_3 = |\partial\mathbf{r}/\partial z| = 1,$$

$$ds^2 = dR^2 + R^2 d\psi^2 + dz^2, \qquad dV = R\,dR\,d\psi\,dz.$$

(ii) *Spherical polar coordinates* (Fig. 5.2):

$$u_1 = r, \quad u_2 = \theta, \quad u_3 = \psi,$$

$$\mathbf{r} = r\sin\theta\cos\psi\,\mathbf{i} + r\sin\theta\sin\psi\,\mathbf{j} + r\cos\theta\mathbf{k},$$

$$h_1 = |\partial\mathbf{r}/\partial r| = 1, \quad h_2 = |\partial\mathbf{r}/\partial\theta| = r, \quad h_3 = |\partial\mathbf{r}/\partial\psi| = r\sin\theta,$$

$$ds^2 = dr^2 + r^2 d\theta^2 + r^2\sin^2\theta\,d\psi^2, \qquad dV = r^2\sin\theta\,dr\,d\theta\,d\psi.$$

Problem 5.1 Express the vector $\mathbf{F} = 3y\mathbf{i} + \mathbf{j} - z^2\mathbf{k}$ in terms of cylindrical polar coordinates and the unit vectors $\mathbf{e}_1 = \hat{\mathbf{R}}$, $\mathbf{e}_2 = \hat{\boldsymbol{\psi}}$, $\mathbf{e}_3 = \hat{\mathbf{z}}$.

Solution. By Fig. 5.1, resolving $\hat{\mathbf{R}}$, $\hat{\boldsymbol{\psi}}$, $\hat{\mathbf{z}}$ in the directions $\mathbf{i}$, $\mathbf{j}$, $\mathbf{k}$:

$$\hat{\mathbf{R}} = \cos\psi\mathbf{i} + \sin\psi\,\mathbf{j}, \quad \hat{\boldsymbol{\psi}} = -\sin\psi\,\mathbf{i} + \cos\psi\,\mathbf{j}, \quad \hat{\mathbf{z}} = \mathbf{k}. \tag{5.9}$$

Alternatively, we may obtain these relations by forming the unit vectors in the directions $\partial\mathbf{r}/\partial R$, $\partial\mathbf{r}/\partial\psi$, $\partial\mathbf{r}/\partial z$, respectively, using (i). (See Problem 5.3) Solving (5.8), we get the inverse relations

$$\mathbf{i} = \cos\psi\,\hat{\mathbf{R}} - \sin\psi\,\hat{\boldsymbol{\psi}}, \quad \mathbf{j} = \sin\psi\,\hat{\mathbf{R}} + \cos\psi\,\hat{\boldsymbol{\psi}}, \quad \mathbf{k} = \hat{\mathbf{z}}. \tag{5.10}$$

Substituting for $\mathbf{i}$, $\mathbf{j}$, $\mathbf{k}$ in $\mathbf{F}$, and putting $y = R\sin\psi$, we get on rearrangement

$$\mathbf{F} = \sin\psi(3R\cos\psi + 1)\hat{\mathbf{R}} + (\cos\psi - 3R\sin^2\psi)\hat{\boldsymbol{\psi}} - z^2\hat{\mathbf{z}}. \qquad \square$$

Problem 5.2 Express the vector $\mathbf{F} = y\mathbf{i} - x\mathbf{j} + 2z\mathbf{k}$ in terms of spherical polar coordinates and the unit vectors $\mathbf{e}_1 = \hat{\mathbf{r}}$, $\mathbf{e}_2 = \hat{\boldsymbol{\theta}}$, $\mathbf{e}_3 = \hat{\boldsymbol{\psi}}$.

Solution. By Fig. 5.2, we get by resolving (or calculation; see Problem 5.3),

$$\begin{aligned} \hat{\mathbf{r}} &= \sin\theta(\cos\psi\,\mathbf{i} + \sin\psi\,\mathbf{j}) + \cos\theta\,\mathbf{k}, \\ \hat{\boldsymbol{\theta}} &= \cos\theta(\cos\psi\,\mathbf{i} + \sin\psi\,\mathbf{j}) - \sin\theta\,\mathbf{k}, \\ \hat{\boldsymbol{\psi}} &= -\sin\psi\,\mathbf{i} + \cos\psi\,\mathbf{j}, \end{aligned} \tag{5.11}$$

since the bracketed vector is the unit vector in the direction of OA. Solving,

$$\begin{aligned} \mathbf{i} &= \cos\psi(\sin\theta\,\hat{\mathbf{r}} + \cos\theta\,\hat{\boldsymbol{\theta}}), \\ \mathbf{j} &= \sin\psi(\sin\theta\,\hat{\mathbf{r}} + \cos\theta\,\hat{\boldsymbol{\theta}}), \\ \mathbf{k} &= \cos\theta\,\hat{\mathbf{r}} - \sin\theta\,\hat{\boldsymbol{\theta}}. \end{aligned} \tag{5.12}$$

Substituting for **i**, **j**, **k** in the given expression for **F**, and putting $x = r \sin\theta \cos\psi$, etc., using (ii), we get after rearranging,

$$\mathbf{F} = 2r\cos^2\theta\,\hat{\mathbf{r}} - 2r\cos\theta\sin\theta\,\hat{\boldsymbol{\theta}} - r\sin\theta\,\hat{\boldsymbol{\psi}}. \qquad \square$$

Problem 5.3 Find the unit vectors $\mathbf{e}_1, \mathbf{e}_2, \mathbf{e}_3$ and the scale factors h_1, h_2, h_3 for the *parabolic cylinder coordinates* (u, v, z):

$$x = \tfrac{1}{2}(u^2 - v^2), \quad y = uv, \quad z = z, \tag{5.13}$$
$$-\infty < u < \infty, \quad 0 \leqslant v < \infty, \quad -\infty < z < \infty.$$

Show that the system is orthogonal, and give the volume element in these coordinates.

Solution. We have $u_1 = u$, $u_2 = v$, $u_3 = z$, and

$$\mathbf{r} = x\mathbf{i} + y\mathbf{j} + z\mathbf{k} = \tfrac{1}{2}(u^2 - v^2)\mathbf{i} + uv\mathbf{j} + z\mathbf{k},$$

$$\frac{\partial \mathbf{r}}{\partial u} = u\mathbf{i} + v\mathbf{j} = h_1\,\mathbf{e}_1, \quad \frac{\partial \mathbf{r}}{\partial v} = -v\mathbf{i} + u\mathbf{j} = h_2\,\mathbf{e}_2, \quad \frac{\partial \mathbf{r}}{\partial z} = \mathbf{k} = h_3\,\mathbf{e}_3, \tag{5.14}$$

by (5.4) and two similar equations. Dividing by magnitudes,

$$\mathbf{e}_1 = \frac{u\mathbf{i} + v\mathbf{j}}{\sqrt{(u^2 + v^2)}}, \quad \mathbf{e}_2 = \frac{-v\mathbf{i} + u\mathbf{j}}{\sqrt{(u^2 + v^2)}}, \quad \mathbf{e}_3 = \mathbf{k}.$$

Also, equating magnitudes on each side in (5.14),

$$h_1 = \sqrt{(u^2 + v^2)}, \quad h_2 = \sqrt{(u^2 + v^2)}, \quad h_3 = 1.$$

By inspection, $\mathbf{e}_1 . \mathbf{e}_2 = \mathbf{e}_2 . \mathbf{e}_3 = \mathbf{e}_3 . \mathbf{e}_1 = 0$, and so the system is orthogonal. Therefore, the volume element is

$$dV = h_1\,h_2\,h_3\,du\,dv\,dz = (u^2 + v^2)\,du\,dv\,dz. \qquad \square$$

Problem 5.4 Show analytically that if u_1, u_2, u_3 are orthogonal curvilinear coordinates defined by the transformation

$$x = x(u_1, u_2, u_3), \quad y = y(u_1, u_2, u_3), \quad z = z(u_1, u_2, u_3), \tag{5.15}$$

and J denotes the Jacobian

$$J = \begin{vmatrix} \partial x/\partial u_1 & \partial x/\partial u_2 & \partial x/\partial u_3 \\ \partial y/\partial u_1 & \partial y/\partial u_2 & \partial y/\partial u_3 \\ \partial z/\partial u_1 & \partial z/\partial u_2 & \partial z/\partial u_3 \end{vmatrix}, \tag{5.16}$$

then $|J| = h_1\,h_2\,h_3$. Give an interpretation.

Solution. The columns of the determinant (5.16) are the rectangular cartesian components of the vectors $\partial\mathbf{r}/\partial u_1$, $\partial\mathbf{r}/\partial u_2$, $\partial\mathbf{r}/\partial u_3$ respectively, and therefore (5.16) is the determinant form for the scalar triple product of these three vectors. Thus,

$$J = \frac{\partial \mathbf{r}}{\partial u_1} \cdot \left(\frac{\partial \mathbf{r}}{\partial u_2} \wedge \frac{\partial \mathbf{r}}{\partial u_3} \right) = h_1 \mathbf{e}_1 . (h_2 \mathbf{e}_2 \wedge h_3 \mathbf{e}_3)$$

$$= \pm h_1 h_2 h_3,$$

since $\mathbf{e}_1$, $\mathbf{e}_2$, $\mathbf{e}_3$ are mutually orthogonal unit vectors. If the system is right-handed, i.e. if $\mathbf{e}_1$, $\mathbf{e}_2$, $\mathbf{e}_3$, in that order, form a right-handed triad at each point, the positive sign applies. If the system is left-handed the negative sign applies. In either case, $|J| = h_1 h_2 h_3$.

If the transformation (5.15) is performed on a triple integral over a region of xyz space, the volume element $dx\, dy\, dz$ is replaced by $dV = |J|\, du_1\, du_2\, du_3$. But, by (5.8), the volume element in the curvilinear coordinate system is also $h_1 h_2 h_3\, du_1\, du_2\, du_3$. Hence $|J|$ and the product of the h's are equivalent expressions for the ratio of corresponding volume elements in xyz space and $u_1 u_2 u_3$ space. □

Problem 5.5 For the *parabolic coordinates* (u, v, ψ):

$$x = uv \cos\psi, \quad y = uv \sin\psi, \quad z = \tfrac{1}{2}(u^2 - v^2), \tag{5.17}$$

$$0 \leqslant u < \infty, \quad 0 \leqslant v < \infty, \quad 0 \leqslant \psi < 2\pi,$$

(i) find the scale factors; (ii) describe the forms of the coordinate surfaces, and give the area elements on them.

Solution. (i) By (5.17),

$$\begin{aligned} \partial\mathbf{r}/\partial u &= v(\cos\psi\, \mathbf{i} + \sin\psi\, \mathbf{j}) + u\mathbf{k}, \\ \partial\mathbf{r}/\partial v &= u(\cos\psi\, \mathbf{i} + \sin\psi\, \mathbf{j}) - v\mathbf{k}, \\ \partial\mathbf{r}/\partial\psi &= uv(-\sin\psi\, \mathbf{i} + \cos\psi\, \mathbf{j}). \end{aligned} \tag{5.18}$$

Taking magnitudes,

$$h_1 = (u^2 + v^2)^{\frac{1}{2}}, \quad h_2 = (u^2 + v^2)^{\frac{1}{2}}, \quad h_3 = uv.$$

(ii) Eliminating the pairs of variables v, ψ; ψ, u; u, v from (5.17) in turn gives

$$2u^2 z = u^4 - (x^2 + y^2), \quad 2v^2 z = x^2 + y^2 - v^4, \quad y = x \tan\psi. \tag{5.19}$$

By the first of (5.19), the coordinate surfaces $u = u_0 =$ constant are the paraboloids of revolution obtained by rotating the confocal parabolas $y^2 = -2u_0^2(z - \frac{1}{2}u_0^2)$, $x = 0$, about Oz. These paraboloids have vertex upwards. The second of (5.19) shows that the coordinate surfaces $v = v_0 =$ constant are also confocal paraboloids of revolution about Oz, but with vertices downwards. In both cases the origin is common focus. Finally, the last equation shows that the surfaces $\psi = \psi_0 =$ constant are planes through the z-axis.

The area elements on these respective surfaces are

$$dS_1 = h_2 h_3\, dv\, d\psi = uv(u^2+v^2)^{\frac{1}{2}} dv\, d\psi,$$
$$dS_2 = h_3 h_1\, d\psi\, du = uv(u^2+v^2)^{\frac{1}{2}} d\psi\, du,$$
$$dS_3 = h_1 h_2\, du\, dv = (u^2+v^2)\, du\, dv. \qquad \square$$

5.2 Nabla Operations In a right-handed orthogonal curvilinear system (u_1, u_2, u_3), where the unit tangent vectors to coordinate lines are $\mathbf{e}_1, \mathbf{e}_2, \mathbf{e}_3$ respectively, and the scale factors are h_1, h_2, h_3, the following formulae apply. They may be obtained by performing a coordinate transformation on the corresponding formulae in rectangular cartesian coordinates, or by a variety of shorter methods described in standard texts.

$$\text{grad}\,\phi = \frac{1}{h_1}\frac{\partial\phi}{\partial u_1}\mathbf{e}_1 + \frac{1}{h_2}\frac{\partial\phi}{\partial u_2}\mathbf{e}_2 + \frac{1}{h_3}\frac{\partial\phi}{\partial u_3}\mathbf{e}_3, \tag{5.20}$$

$$\text{div}\,\mathbf{F} = \frac{1}{h_1 h_2 h_3}\left[\frac{\partial}{\partial u_1}(h_2 h_3 F_1) + \frac{\partial}{\partial u_2}(h_3 h_1 F_2) + \frac{\partial}{\partial u_3}(h_1 h_2 F_3)\right], \tag{5.21}$$

$$\nabla^2\phi = \frac{1}{h_1 h_2 h_3}\left[\frac{\partial}{\partial u_1}\left(\frac{h_2 h_3}{h_1}\frac{\partial\phi}{\partial u_1}\right) + \frac{\partial}{\partial u_2}\left(\frac{h_3 h_1}{h_2}\frac{\partial\phi}{\partial u_2}\right) + \frac{\partial}{\partial u_3}\left(\frac{h_1 h_2}{h_3}\frac{\partial\phi}{\partial u_3}\right)\right], \tag{5.22}$$

$$\text{curl}\,\mathbf{F} = \begin{vmatrix} h_1\mathbf{e}_1 & h_2\mathbf{e}_2 & h_3\mathbf{e}_3 \\ \partial/\partial u_1 & \partial/\partial u_2 & \partial/\partial u_3 \\ h_1 F_1 & h_2 F_2 & h_3 F_3 \end{vmatrix}. \tag{5.23}$$

Here, ϕ is a continuous scalar function and $\mathbf{F} = F_1\mathbf{e}_1 + F_2\mathbf{e}_2 + F_3\mathbf{e}_3$ is a continuous vector function, each possessing continuous partial derivatives of necessary orders.

In *cylindrical polar coordinates*, these become

$$\text{grad}\,\phi = \frac{\partial\phi}{\partial R}\hat{\mathbf{R}} + \frac{1}{R}\frac{\partial\phi}{\partial\psi}\hat{\boldsymbol{\psi}} + \frac{\partial\phi}{\partial z}\hat{\mathbf{z}}, \tag{5.24}$$

$$\text{div}\,\mathbf{F} = \frac{1}{R}\frac{\partial}{\partial R}(RF_R) + \frac{1}{R}\frac{\partial F_\psi}{\partial\psi} + \frac{\partial F_z}{\partial z}, \tag{5.25}$$

$$\nabla^2\phi = \frac{1}{R}\frac{\partial}{\partial R}\left(R\frac{\partial\phi}{\partial R}\right) + \frac{1}{R^2}\frac{\partial^2\phi}{\partial\psi^2} + \frac{\partial^2\phi}{\partial z^2}, \tag{5.26}$$

$$\operatorname{curl}\mathbf{F} = \frac{1}{R}\begin{vmatrix} \hat{\mathbf{R}} & R\hat{\boldsymbol{\psi}} & \hat{\mathbf{z}} \\ \partial/\partial R & \partial/\partial\psi & \partial/\partial z \\ F_R & RF_\psi & F_z \end{vmatrix}. \tag{5.27}$$

In *spherical polar coordinates*,

$$\operatorname{grad}\phi = \frac{\partial\phi}{\partial r}\hat{\mathbf{r}} + \frac{1}{r}\frac{\partial\phi}{\partial\theta}\hat{\boldsymbol{\theta}} + \frac{1}{r\sin\theta}\frac{\partial\phi}{\partial\psi}\hat{\boldsymbol{\psi}}, \tag{5.28}$$

$$\operatorname{div}\mathbf{F} = \frac{1}{r^2}\frac{\partial}{\partial r}(r^2F_r) + \frac{1}{r\sin\theta}\frac{\partial}{\partial\theta}(\sin\theta\, F_\theta) + \frac{1}{r\sin\theta}\frac{\partial F_\psi}{\partial\psi}, \tag{5.29}$$

$$\nabla^2\phi = \frac{1}{r^2}\frac{\partial}{\partial r}\left(r^2\frac{\partial\phi}{\partial r}\right) + \frac{1}{r^2\sin\theta}\frac{\partial}{\partial\theta}\left(\sin\theta\frac{\partial\phi}{\partial\theta}\right) + \frac{1}{r^2\sin^2\theta}\frac{\partial^2\phi}{\partial\psi^2}, \tag{5.30}$$

$$\operatorname{curl}\mathbf{F} = \frac{1}{r^2\sin\theta}\begin{vmatrix} \hat{\mathbf{r}} & r\hat{\boldsymbol{\theta}} & r\sin\theta\hat{\boldsymbol{\psi}} \\ \partial/\partial r & \partial/\partial\theta & \partial/\partial\psi \\ F_r & rF_\theta & r\sin\theta\, F_\psi \end{vmatrix}. \tag{5.31}$$

Problem 5.6 Find $\operatorname{grad}\phi$, where $\phi = r^3\cos\theta$ in spherical polar coordinates.

Solution. We have

$$\partial\phi/\partial r = 3r^2\cos\theta, \quad \partial\phi/\partial\theta = -r^3\sin\theta, \quad \partial\phi/\partial\psi = 0.$$

Hence by (5.28),

$$\operatorname{grad}\phi = 3r^2\cos\theta\,\hat{\mathbf{r}} - r^2\sin\theta\,\hat{\boldsymbol{\theta}}. \qquad \square$$

Problem 5.7 If $\mathbf{q} = \cos\psi\,\hat{\mathbf{R}} + R\sin\psi\,\hat{\boldsymbol{\psi}}$, in cylindrical polar coordinates, find the locus of points at which $\mathbf{q}\wedge\operatorname{curl}\mathbf{q} = 0$.

Solution. By (5.27),

$$\operatorname{curl}\mathbf{q} = \frac{1}{R}\begin{vmatrix} \hat{\mathbf{R}} & R\hat{\boldsymbol{\psi}} & \hat{\mathbf{z}} \\ \partial/\partial R & \partial/\partial\psi & \partial/\partial z \\ \cos\psi & R^2\sin\psi & 0 \end{vmatrix}$$

$$= R^{-1}[2R\sin\psi - (-\sin\psi)]\hat{\mathbf{z}} = (2+R^{-1})\sin\psi\,\hat{\mathbf{z}}, \quad (R \neq 0).$$

Therefore,

$$\mathbf{q}\wedge\operatorname{curl}\mathbf{q} = \begin{vmatrix} \hat{\mathbf{R}} & \hat{\boldsymbol{\psi}} & \hat{\mathbf{z}} \\ \cos\psi & R\sin\psi & 0 \\ 0 & 0 & (2+R^{-1})\sin\psi \end{vmatrix}$$

$$= (2R+1)\sin^2\psi\,\hat{\mathbf{R}} - (2+R^{-1})\cos\psi\sin\psi\,\hat{\boldsymbol{\psi}}, \quad (R \neq 0).$$

The required locus is given by the equation $\sin\psi = 0$, i.e. $\psi = 0$ or $\psi = \pi$, which corresponds to the plane $y = 0$.

Note that the determinant form for vector product may be assumed in any right-handed system of orthogonal curvilinear coordinates, since the unit triad $(\mathbf{e}_1, \mathbf{e}_2, \mathbf{e}_3)$ at every point has the same algebraic properties as $(\mathbf{i}, \mathbf{j}, \mathbf{k})$. □

Problem 5.8 Determine all solutions of the form $\phi(r, t)$ of the wave equation

$$\nabla^2\phi - \frac{1}{c^2}\frac{\partial^2\phi}{\partial t^2} = 0, \tag{5.32}$$

(where c is a constant) in spherical polar coordinates.

Solution. By (5.30), since ϕ is to be assumed independent of θ and ψ, the wave equation takes the form

$$\frac{\partial^2\phi}{\partial r^2} + \frac{2}{r}\frac{\partial\phi}{\partial r} - \frac{1}{c^2}\frac{\partial^2\phi}{\partial t^2} = 0,$$

or, multiplying by r,

$$\frac{\partial^2}{\partial r^2}(r\phi) - \frac{1}{c^2}\frac{\partial^2}{\partial t^2}(r\phi) = 0. \tag{5.33}$$

Make the substitution $u = r + ct$, $v = r - ct$. Then

$$\frac{\partial}{\partial r} = \frac{\partial u}{\partial r}\frac{\partial}{\partial u} + \frac{\partial v}{\partial r}\frac{\partial}{\partial v} = \frac{\partial}{\partial u} + \frac{\partial}{\partial v},$$

$$\frac{1}{c}\frac{\partial}{\partial t} = \frac{1}{c}\left(\frac{\partial u}{\partial t}\frac{\partial}{\partial u} + \frac{\partial v}{\partial t}\frac{\partial}{\partial v}\right) = \frac{\partial}{\partial u} - \frac{\partial}{\partial v},$$

so that (5.33) becomes

$$\left[\left(\frac{\partial}{\partial u} + \frac{\partial}{\partial v}\right)^2 - \left(\frac{\partial}{\partial u} - \frac{\partial}{\partial v}\right)^2\right](r\phi) = 4\frac{\partial^2}{\partial u \partial v}(r\phi) = 0.$$

Successive integrations give

$$\frac{\partial}{\partial v}(r\phi) = g(v), \quad r\phi = F(u) + G(v),$$

where F and $G = \int g\, dv$ are arbitrary functions. Hence the required solutions are

$$\phi = r^{-1}[F(r+ct) + G(r-ct)].$$ □

Problem 5.9 Express Laplace's equation $\nabla^2\phi = 0$ in parabolic cylinder coordinates (u, v, z), given by (5.13), and find all solutions of the form $\phi = f(u^2+v^2)$.

Solution. By Problem 5.3, the scale factors are

$$h_1 = h_2 = (u^2+v^2)^{\frac{1}{2}}, \qquad h_3 = 1,$$

and substituting in (5.22) we obtain for Laplace's equation

$$\nabla^2\phi = \frac{1}{u^2+v^2}\left(\frac{\partial^2\phi}{\partial u^2}+\frac{\partial^2\phi}{\partial v^2}\right)+\frac{\partial^2\phi}{\partial z^2} = 0.$$

Substituting $\phi = f(w)$, where $w = u^2+v^2$, we find

$$wf''+f' = 0, \qquad (f' \equiv df/dw)$$

giving on integration

$$wf' = A, \qquad \text{i.e. } f = A\ln w+B,$$

where A and B are arbitrary constants. The required solutions are therefore

$$\phi = A\ln(u^2+v^2)+B. \qquad \square$$

EXERCISES

1. If $\mathbf{F} = R\sin\psi\,\hat{\mathbf{R}}+2R(\cos\psi-\sin\psi)\hat{\boldsymbol{\psi}}$ in cylindrical polar coordinates, find div $\mathbf{F}$, and hence show that the locus of points at which div $\mathbf{F}$ vanishes is the plane $x = 0$.

2. Find curl $\mathbf{F}$, where $\mathbf{F} = r\hat{\mathbf{r}}+r^2\sin\theta\,\hat{\boldsymbol{\theta}}+r^2\cos\theta\,\hat{\boldsymbol{\psi}}$ in spherical polar coordinates.

3. Show that *prolate spheroidal coordinates* (u, v, ψ):

$$x = a\sinh u\sin v\cos\psi, \quad y = a\sinh u\sin v\sin\psi, \quad z = a\cosh u\cos v,$$

$$0 \leqslant u < \infty, \quad 0 \leqslant v \leqslant \pi, \quad 0 \leqslant \psi < 2\pi,$$

are orthogonal. Find (i) the scale factors, (ii) the area elements on coordinate surfaces.

4. Express Laplace's equation $\nabla^2\phi = 0$ in *elliptic cylinder coordinates* (u, v, z):

$$x = c\cosh u\cos v, \quad y = c\sinh u\sin v, \quad z = z,$$

$$0 \leqslant u < \infty, \quad 0 \leqslant v < 2\pi, \quad -\infty < z < \infty,$$

and find the general solution of the form $\phi = f(u+v)$.

5. Express the vector $\mathbf{i}+y\mathbf{k}$ in parabolic coordinates (see Problem 5.5).

6. Find the volume element in (i) elliptic cylinder coordinates, (ii) parabolic coordinates, (see Exercise 4 and Problem 5.5).

Chapter 6

Cartesian Tensors

6.1 Vectors and Orthogonal Transformations Let **A** be a given vector, which we shall suppose is defined without reference to any particular coordinate system. To fix our ideas, we may imagine **A** to be defined as the directed line-segment **PQ**, where P and Q are particles in a certain body. If $Oxyz$ and $Ox'y'z'$ are two sets of rectangular cartesian axes sharing a common origin (Fig. 6.1), then the components of **A** in the first system will be a triple (A_x, A_y, A_z) and the components in the second system

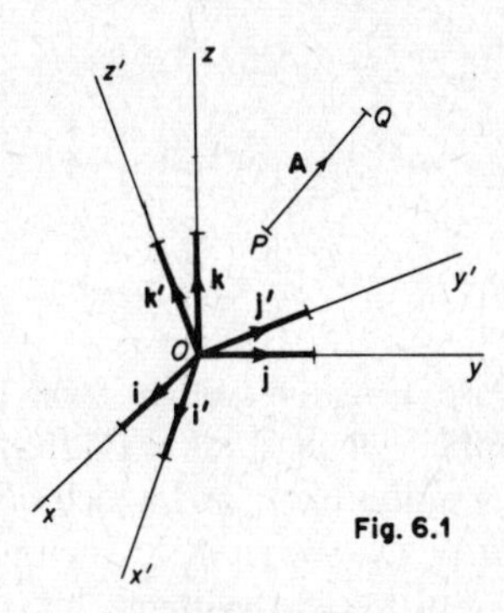

Fig. 6.1

will be a triple (A'_x, A'_y, A'_z). Since **PQ** is the same line-segment in each case, there is a definite relationship between the two triples in terms of the relative orientation of the systems of axes, and this relationship, to be determined, forms an integral part of the analytical definition of a vector.

Let **i**, **j**, **k** have their usual meanings for the system x, y, z. The corresponding unit vectors **i**′, **j**′, **k**′ for the second system can be expressed in the component form

$$\begin{aligned} \mathbf{i}' &= a_{11}\,\mathbf{i}+a_{12}\,\mathbf{j}+a_{13}\,\mathbf{k}, \\ \mathbf{j}' &= a_{21}\,\mathbf{i}+a_{22}\,\mathbf{j}+a_{23}\,\mathbf{k}, \\ \mathbf{k}' &= a_{31}\,\mathbf{i}+a_{32}\,\mathbf{j}+a_{33}\,\mathbf{k}. \end{aligned} \tag{6.1}$$

Forming the scalar product of each equation with **i**, **j**, **k**, in turn,

$$a_{11} = \mathbf{i}'.\mathbf{i} = \cos(x', x), \quad a_{12} = \mathbf{i}'.\mathbf{j} = \cos(x', y), \ldots, \quad a_{33} = \mathbf{k}'.\mathbf{k} = \cos(z', z). \tag{6.2}$$

F

Since

$$\mathbf{A} = A_x\mathbf{i}+A_y\mathbf{j}+A_z\mathbf{k} = A'_x\mathbf{i}'+A'_y\mathbf{j}'+A'_z\mathbf{k}',$$

we get by substitution from (6.1)

$$\begin{aligned} A_x &= a_{11}A'_x+a_{21}A'_y+a_{31}A'_z, \\ A_y &= a_{12}A'_x+a_{22}A'_y+a_{32}A'_z, \\ A_z &= a_{13}A'_x+a_{23}A'_y+a_{33}A'_z, \end{aligned} \tag{6.3}$$

which shows, together with (6.2), how the triples are related.

Introducing the *suffix notation* (x_1, x_2, x_3) in place of (x, y, z) allows us to write (6.3) more concisely as

$$\begin{aligned} A_1 &= \sum_{j=1}^{3} a_{j1}A'_j = a_{j1}A'_j, \\ A_2 &= \sum_{j=1}^{3} a_{j2}A'_j = a_{j2}A'_j, \\ A_3 &= \sum_{j=1}^{3} a_{j3}A'_j = a_{j3}A'_j, \end{aligned} \tag{6.4}$$

where on the right we use Einstein's *summation convention*, by which, summation is always implied when a suffix taking values 1, 2, 3 appears *twice* in any expression. A suffix like i or j which appears once is called a *free* suffix and takes each of the values 1, 2, 3 without summation. Thus, the complete set of equations (6.4) is written

$$A_i = a_{ji}A'_j. \tag{6.5}$$

Notice that symbols used to denote general suffixes can be replaced by others, without altering the expression concerned, provided this is done consistently. For example, the equation $A_p = a_{qp}A'_q$ is identical to (6.5).

Interchanging the roles of primed and unprimed symbols shows that we can write analogously to (6.5)

$$A'_i = b_{ji}A_j, \tag{6.6}$$

where by (6.2) we have the nine equations

$$b_{ji} = \cos(x_j, x'_i) = \cos(x'_i, x_j) = a_{ij} .$$

Therefore, (6.6) becomes in terms of the a's

$$A'_i = a_{ij}A_j. \tag{6.7}$$

The transformations (6.5) and (6.7) are equivalent, each being the inverse of the other.

Since for arbitrary A_i,

$$A_i = a_{ji}A'_j = a_{ji}a_{jk}A_k,$$

we must have $$a_{ji}\, a_{jk} = \delta_{ik}, \tag{6.8}$$
where $$\delta_{ik} = 0 \text{ if } i \neq k; \quad \delta_{ik} = 1 \text{ if } i = k.$$

This symbol is the *Kronecker delta*, or the *substitution* symbol, the latter name originating from the fact that multiplying A_k by δ_{ik} has the same effect as substituting A_i for A_k.

Problem 6.1 (i) Write out in full $c_{1i}\, c_{ik} = 0$. Evaluate (ii) δ_{ii}, (iii) $\delta_{ij}\, \delta_{jk}\, A_k$.

Solution. (i) Applying the summation convention and setting $k = 1, 2, 3$, we obtain

$$c_{11}\, c_{11} + c_{12}\, c_{21} + c_{13}\, c_{31} = 0, \quad c_{11}\, c_{12} + c_{12}\, c_{22} + c_{13}\, c_{32} = 0,$$
$$c_{11}\, c_{13} + c_{12}\, c_{23} + c_{13}\, c_{33} = 0.$$

(ii) $\delta_{ii} = \delta_{11} + \delta_{22} + \delta_{33} = 3.$

(iii) The expression is $\delta_{ij}\,(\delta_{jk}\, A_k) = \delta_{ij}\, A_j = A_i$. □

If the coordinates of a point C in space are (x_1, x_2, x_3) in the unprimed system and (x'_1, x'_2, x'_3) in the primed system, then by applying the foregoing results to the vector **OC** we obtain by (6.7), (6.5),

$$x'_i = a_{ij}\, x_j, \tag{6.9}$$

and the inverse transformation

$$x_i = a_{ji}\, x'_j. \tag{6.10}$$

A coordinate transformation of the form (6.9) for which (6.8) applies,

$$a_{ji}\, a_{jk} = \delta_{ik},$$

is termed *orthogonal.* Of these nine equations, for different combinations of i and k, only six are independent. For,

$$\begin{aligned} a_{ji}\, a_{jk} &= a_{1i}\, a_{1k} + a_{2i}\, a_{2k} + a_{3i}\, a_{3k} \\ &= a_{1k}\, a_{1i} + a_{2k}\, a_{2i} + a_{3k}\, a_{3i} \\ &= a_{jk}\, a_{ji}, \end{aligned}$$

which shows that if (6.8) holds for $i < k$, then it automatically holds for $i > k$.

Problem 6.2 Show that the transformation a_{ij} given by the matrix

$$(a_{ij}) = \begin{pmatrix} \frac{1}{2} & 0 & \frac{1}{2}\sqrt{3} \\ 0 & 1 & 0 \\ -\frac{1}{2}\sqrt{3} & 0 & \frac{1}{2} \end{pmatrix}$$

is orthogonal.

Solution. If T denotes the matrix (a_{ij}), and $\tilde{T}$ its transpose. i.e. $(\tilde{T})_{ij} = (T)_{ji}$, then the matrix form of (6.8) is

$$\tilde{T}T = I,$$

where I is the 3×3 identity matrix. For the case in question,

$$\tilde{T}T = \begin{pmatrix} \frac{1}{2} & 0 & -\frac{1}{2}\sqrt{3} \\ 0 & 1 & 0 \\ \frac{1}{2}\sqrt{3} & 0 & \frac{1}{2} \end{pmatrix} \begin{pmatrix} \frac{1}{2} & 0 & \frac{1}{2}\sqrt{3} \\ 0 & 1 & 0 \\ -\frac{1}{2}\sqrt{3} & 0 & \frac{1}{2} \end{pmatrix}$$

$$= \begin{pmatrix} 1 & 0 & 0 \\ 0 & 1 & 0 \\ 0 & 0 & 1 \end{pmatrix} = I,$$

and so the transformation a_{ij} is orthogonal. □

Note that $\tilde{T} = T^{-1}$, and so we also have for an orthogonal transformation $T\tilde{T} = I$, or in suffix notation $a_{ij}\,a_{kj} = \delta_{ik}$, which is equivalent to (6.8).

Problem 6.3 Show that (i) for any orthogonal transformation, $\det(a_{ij}) = \pm 1$, and (ii) for a rotation of cartesian axes, $\det(a_{ij}) = +1$.

Solution. (i) As in Problem 6.2 we have $\tilde{T}T = I$, where $\tilde{T}$ is the transpose of the matrix $T = (a_{ij})$, when a_{ij} is orthogonal. By the matrix rule for multiplying determinants, since $|\tilde{T}| = |T|$, we have $|T|^2 = 1$, and so $|T| = \pm 1$.

(ii) Observe first that the 'zero' rotation corresponds to the identity transformation

$$x'_i = x_i = \delta_{ij}\,x_j,$$

for which the transformation matrix is δ_{ij}, and that $\det(\delta_{ij}) = +1$. Since the elements of a_{ij} are, by (6.2), continuous functions of the angles between the axes in the x_i and x'_i systems, it follows by continuity that for any rotation of axes through an *infinitesimal* angle we must have $\det(a_{ij}) = +1$, rather than -1. But a finite rotation with transformation matrix T may be performed as a succession of n infinitesimal transformations $T_1, T_2, \ldots, T_n$, where n is large. Hence,

$$T = T_n\,T_{n-1}\ldots T_1,$$

and taking determinants gives the result. □

The case $|T| = -1$ corresponds to a transformation from a right-handed set of axes to a left-handed set, or vice versa, and arises in particular in a reflection of one or all three axes.

Problem 6.4 Give a_{ij} corresponding to a positive rotation of axes through an angle $\pi/4$ about the x_3-axis.

Solution. The rotation leaves x_3 unchanged, i.e. $x'_3 = x_3$. By a well-known formula, a rotation of axes through an angle θ in the x_1x_2-plane is given by

$$x'_1 = x_1 \cos\theta + x_2 \sin\theta, \quad x'_2 = -x_1 \sin\theta + x_2 \cos\theta.$$

With $\theta = \pi/4$, this is to apply in every plane $x_3 =$ constant, and combining the two results we get for the required transformation $x'_i = a_{ij}x_j$, where

$$a_{ij} = \begin{pmatrix} 1/\sqrt{2} & 1/\sqrt{2} & 0 \\ -1/\sqrt{2} & 1/\sqrt{2} & 0 \\ 0 & 0 & 1 \end{pmatrix}.$$

□

Problem 6.5 (i) A vector $\mathbf{p}$ undergoes positive rotation through an angle θ about an axis in the direction of a *unit* vector $\mathbf{l}$. Express the resulting vector in terms of $\mathbf{p}$, θ and $\mathbf{l}$. (ii) Coordinate axes $Ox'_1\,x'_2\,x'_3$ are obtained by rotating the system $Ox_1\,x_2\,x_3$ through an angle θ about an axis in the direction of a unit vector $\mathbf{l}$. Find the matrix a_{ij} of the corresponding coordinate transformation.

Solution. (i) We may suppose the axis to pass through the origin. Let $\mathbf{OP} = \mathbf{p}$ (Fig. 6.2), and let N be the projection of P on the axis. If Q is the point

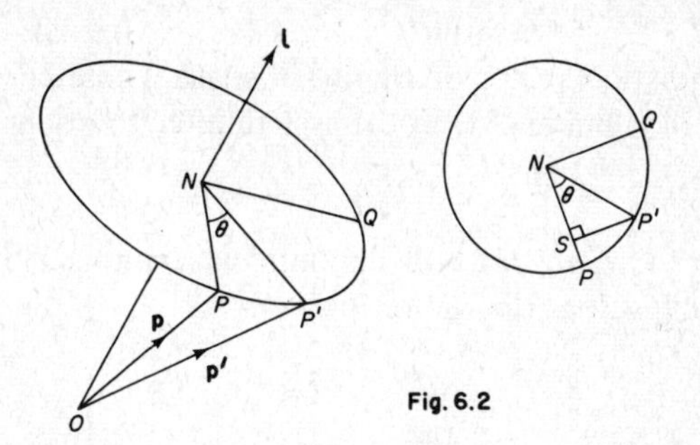

Fig. 6.2

such that $NQ = NP$ and such that $\mathbf{l}$, $\mathbf{NP}$, $\mathbf{NQ}$ form a right-handed system of mutually orthogonal vectors, we have

$$\mathbf{NQ} = \mathbf{l} \wedge \mathbf{p}, \qquad \mathbf{PN} = \mathbf{l} \wedge \mathbf{NQ} = \mathbf{l} \wedge (\mathbf{l} \wedge \mathbf{p}),$$

as may be verified by comparing magnitudes and directions on each side.

If $\mathbf{OP}' = \mathbf{p}'$, then from the diagram

$$\mathbf{PP}' = \mathbf{PS} + \mathbf{SP}' = (1 - \cos\theta)\mathbf{PN} + \sin\theta\,\mathbf{NQ}.$$

Therefore,

$$\mathbf{p}' = \mathbf{p} + \sin\theta\,\mathbf{l} \wedge \mathbf{p} + (1 - \cos\theta)\mathbf{l} \wedge (\mathbf{l} \wedge \mathbf{p}). \tag{6.11}$$

(ii) Let $\mathbf{i}_k$ and $\mathbf{i}'_k$ be unit vectors in the directions of the x_k and x'_k axes, respectively. We shall set $\mathbf{p} = \mathbf{i}_k$ and $\mathbf{p}' = \mathbf{i}'_k$ in (6.11). Writing $\mathbf{l} = l_j\mathbf{i}_j$, we have

$$\mathbf{l} \wedge \mathbf{i}_k = l_j\mathbf{i}_j \wedge \mathbf{i}_k = l_j\varepsilon_{ijk}\mathbf{i}_i,$$

where ε_{ijk} is the *alternating symbol*, which has the value 1 when i, j, k is an even permutation of 1,2,3, the value -1 when i, j, k is an odd permutation of 1,2,3, and the value 0 when two or more of the suffixes are equal. Explicitly, the only non-vanishing cases are

$$\varepsilon_{123} = \varepsilon_{231} = \varepsilon_{312} = 1, \qquad \varepsilon_{132} = \varepsilon_{213} = \varepsilon_{321} = -1.$$

Expanding the vector triple product and using the fact that $\mathbf{l}$ is a *unit* vector,

$$\begin{aligned}\mathbf{l} \wedge (\mathbf{l} \wedge \mathbf{i}_k) &= (\mathbf{l}\,.\,\mathbf{i}_k)\mathbf{l} - (\mathbf{l}\,.\,\mathbf{l})\mathbf{i}_k = (l_j\mathbf{i}_j\,.\,\mathbf{i}_k)\mathbf{l} - \mathbf{i}_k \\ &= (l_j\delta_{jk})l_m\mathbf{i}_m - \mathbf{i}_k = l_k l_m\mathbf{i}_m - \mathbf{i}_k.\end{aligned}$$

Hence, (6.11) gives

$$\begin{aligned}\mathbf{i}'_k &= \mathbf{i}_k + \sin\theta\,\varepsilon_{ijk}\,l_j\mathbf{i}_i + (1-\cos\theta)(l_k\,l_m\,\mathbf{i}_m - \mathbf{i}_k) \\ &= \cos\theta\,\mathbf{i}_k + \sin\theta\,\varepsilon_{ijk}\,l_j\mathbf{i}_i + (1-\cos\theta)l_k\,l_m\,\mathbf{i}_m. \qquad (6.12)\end{aligned}$$

If we put

$$a_{ki} = \cos\theta\,\delta_{ki} + \sin\theta\,\varepsilon_{ijk}\,l_j + (1-\cos\theta)l_k\,l_i, \qquad (6.13)$$

then (6.12) takes the form $\mathbf{i}'_k = a_{ki}\mathbf{i}_i$.

Let $\mathbf{r}$ be the position vector of a point in space whose coordinates are x_i in the original system and x'_i in the system after rotation. By (6.13),

$$\mathbf{r} = x_i\mathbf{i}_i = x'_k\mathbf{i}'_k = x'_k\,a_{ki}\,\mathbf{i}_i,$$

whence we get $x_i = x'_k\,a_{ki}$. Multiplying both sides of this last equation by a_{ji} and using the result at the end of Problem 6.2,

$$x'_i = a_{ij}x_j. \qquad (6.14)$$

The required transformation is given by (6.14) and (6.13). □

Problem 6.6 Give, explicitly, the coordinate transformation corresponding to a rotation of axes through 60° about an axis in the direction of $\mathbf{i}_1 + \mathbf{i}_2$.

Solution. In the notation of the last problem, $\cos\theta = \frac{1}{2}$, $\sin\theta = \frac{1}{2}\sqrt{3}$, $l_1 = l_2 = 1/\sqrt{2}$, $l_3 = 0$. Therefore, by (6.13),

$$a_{11} = \cos\theta + (1-\cos\theta)l_1^2 = \tfrac{3}{4},$$

$$a_{12} = \sin\theta\,l_3 + (1-\cos\theta)l_1\,l_2 = \tfrac{1}{4},$$

etc., the remaining elements being obtained in like fashion. We find

$$(a_{ij}) = \tfrac{1}{4}\begin{pmatrix} 3 & 1 & -\sqrt{6} \\ 1 & 3 & \sqrt{6} \\ \sqrt{6} & -\sqrt{6} & 2 \end{pmatrix}$$

and by (6.14),

$$\begin{aligned} x_1' &= \tfrac{1}{4}(3x_1 + x_2 - \sqrt{6}x_3), \\ x_2' &= \tfrac{1}{4}(x_1 + 3x_2 + \sqrt{6}x_3), \\ x_3' &= \tfrac{1}{4}(\sqrt{6}x_1 - \sqrt{6}x_2 + 2x_3). \end{aligned}$$

□

6.2 Cartesian Tensors In the remainder of this chapter right-handed systems of axes only will be considered. Let x_i be a chosen coordinate system, and let x_i' be any second system, related to the first in accordance with an orthogonal transformation (6.14). If $u_i = (u_1, u_2, u_3)$ is any triple of numbers defined in the x_i system, we say that u_i are the components of a *cartesian tensor of rank* 1 if corresponding triples $u_i' = (u_1', u_2', u_3')$ are defined in every x_i' system and if

$$u_i' = a_{ij} u_j, \tag{6.15}$$

that is, the u_i transform in the same way as the coordinates of a fixed point. Thus, a vector is the same as a cartesian tensor of rank 1.

A pair of vectors u_i, v_i may be combined to form their 'outer product' $u_i v_j$, which is the array of nine numbers

$$(u_i v_j) = \begin{pmatrix} u_1 v_1 & u_1 v_2 & u_1 v_3 \\ u_2 v_1 & u_2 v_2 & u_2 v_3 \\ u_3 v_1 & u_3 v_2 & u_3 v_3 \end{pmatrix}. \tag{6.16}$$

The outer product transforms according to the rule

$$u_i' v_j' = (a_{ik} u_k)(a_{jl} v_l) = a_{ik} a_{jl} u_k v_l.$$

Any set of nine numbers t_{ij} defined in the x_i system (not necessarily the outer product of two vectors) are the components of a *cartesian tensor of rank* 2 if corresponding numbers t_{ij}' are defined in each x_i' system, such that

$$t_{ij}' = a_{ik} a_{jl} t_{kl},$$

where a_{ij} is the transformation matrix.

In general, a cartesian tensor of rank n is a set of 3^n numbers $t_{i_1 i_2 \cdots i_n}$ which transforms according to the law

$$t'_{i_1 i_2 \cdots i_n} = a_{i_1 j_1} a_{i_2 j_2} \cdots a_{i_n j_n} t_{j_1 j_2 \cdots j_n},$$

on changing from the x_i to the x_i' system.

A *scalar* is a single number s which takes the same value in every cartesian system; it is called a *cartesian tensor of rank* 0. For example, the scalar product of two vectors u_i and v_i is the number $u_i v_i$, which *is* a scalar since

$$u'_i v'_i = a_{ik} a_{il} u_k v_l = \delta_{kl} u_k v_l = u_k v_k.$$

Cartesian tensors of the same rank may be added or subtracted by adding or subtracting corresponding components, and a tensor may be multiplied by a scalar by multiplying each of its components by the scalar.

If two free indices of a tensor are put equal, the implied summation reduces the number of free indices by 2. This process is called *contraction*, and the resulting quantity may be shown always to be a tensor. That is, contraction of a cartesian tensor of rank $n(> 2)$, with respect to any pair of indices, always produces a cartesian tensor of rank $n-2$.

Problem 6.7 If t_{ijkl} is a cartesian tensor of rank 4, show that $u_{jk} = t_{ijki}$ is a cartesian tensor of rank 2.

Solution. On transforming from the x_i system to the x'_i system, u_{jk} becomes (using the orthogonality of a_{ij}),

$$\begin{aligned} u'_{jk} &= t'_{ijki} = a_{ip} a_{jq} a_{kr} a_{is} t_{pqrs} = a_{jq} a_{kr} (a_{ip} a_{is}) t_{pqrs} \\ &= a_{jq} a_{kr} \delta_{ps} t_{pqrs} = a_{jl} a_{kr} t_{pqrp} \\ &= a_{jq} a_{kr} u_{qr}, \end{aligned}$$

which proves the result. □

Problem 6.8 (i) Show that the Kronecker delta δ_{ij} is a cartesian tensor of rank 2. (ii) Show that the permutation symbol ε_{ijk} is a cartesian tensor of rank 3 under proper orthogonal transformations (i.e. those with $\det(a_{ij}) = +1$).

Solution. (i) The problem implies that the definition of δ_{ij}, as given earlier, is to apply in every cartesian system. We require to show that this is consistent with the tensor law of transformation. Let d_{ij} be a cartesian tensor, and let $d_{ij} = \delta_{ij}$ in the x_i system. In any other system x'_i,

$$\begin{aligned} d'_{ij} &= a_{ik} a_{jl} d_{kl} = a_{ik} a_{jl} \delta_{kl} \\ &= a_{ik} a_{jk} = \delta_{ij} = \delta'_{ij}, \quad \text{(given)}, \end{aligned}$$

since a_{ij} is orthogonal. Therefore d_{ij} and δ_{ij} are identical in every system, and since the former is a tensor the result follows.

(ii) Since $\varepsilon'_{ijk} = \varepsilon_{ijk}$, we need to show that

$$a_{il} a_{jm} a_{kn} \varepsilon_{lmn} = \varepsilon_{ijk}. \tag{6.17}$$

Expanded, the left-hand side is

$$a_{i1}\,a_{j2}\,a_{k3}-a_{i1}\,a_{j3}\,a_{k2}+a_{i2}\,a_{j3}\,a_{k1}-a_{i2}\,a_{j1}\,a_{k3}+a_{i3}\,a_{j1}\,a_{k2}-a_{i3}\,a_{j2}\,a_{k1},$$

which is also the expansion of the determinant

$$\begin{vmatrix} a_{i1} & a_{i2} & a_{i3} \\ a_{j1} & a_{j2} & a_{j3} \\ a_{k1} & a_{k2} & a_{k3} \end{vmatrix}. \tag{6.18}$$

If two of the suffixes i, j, k are equal, the determinant has two identical rows, and so vanishes. If i, j, k are distinct and form an even permutation of 1,2,3, then an even permutation of rows brings (6.18) (without changing its value) into the form det (a_{ij}), whose value is 1. If i, j, k are distinct and form an odd permutation of 1,2,3, then an odd permutation of rows brings (6.18) into the form det (a_{ij}), while at the same time introducing a sign change. It follows that in all cases the value of (6.18) is ε_{ijk}, which proves (6.17) and the stated result. □

Problem 6.9 Show that

$$\text{(i)}\ \varepsilon_{ijm}\,\varepsilon_{ijn} = 2\delta_{mn}, \quad \text{(ii)}\ \varepsilon_{ijk}\,\varepsilon_{ipq} = \delta_{jp}\,\delta_{kq}-\delta_{jq}\,\delta_{kp}.$$

Solution. (i) The component ε_{ijm} vanishes unless i, j, m are all different and likewise ε_{ijn} vanishes unless i, j, n are all different. Since only three of the numbers i, j, m, n can be distinct, the left-hand side vanishes for $m \neq n$. When $m = n$, only two non-zero terms occur in the summation over i and j; in one such term both ε symbols have the value -1, and in the other both have the value $+1$. Hence, for $m = n$, the sum of the terms is 2, which proves the result.

(ii) Here, non-zero terms on the left occur only if p and q are distinct and if we also have $j = p$, $k = q$, or $j = q$, $k = p$. In the former case, there is just one non-zero term, corresponding to the unique value of i which differs from both p and q. The factors in this term are equal and have the product 1. In the latter case there is again just one non-zero term, but the factors ε_{iqp} and ε_{ipq} have opposite signs, and their product is -1. It follows that if p and q are unequal, the left-hand side has the value 1 if $j = p$, $k = q$, and the value -1 if $j = q$, $k = p$. The value is zero in all other cases. By inspection, the right-hand side takes precisely the same values in these cases, and so (ii) is verified. □

6.3 Cartesian Tensor Fields A tensor defined at each point of a region forms a *tensor field*. A *scalar field* is a tensor field of rank 0, i.e. a scalar which has a value at each point of the region, the value being independent of the choice of coordinate system. Thus, if $\phi(x_1, x_2, x_3)$ is the value in the

x_i system and $\phi'(x_2', x_2', x_3')$ is the value in the x_i' system, then if the coordinates refer to the same point in the region we have

$$\phi'(x_1', x_2', x_3') = \phi(x_1, x_2, x_3).$$

A *vector field* is a tensor field of rank 1. We are concerned only with *cartesian* tensor and scalar fields.

Problem 6.10 Show that the components of the gradient of a scalar field form a vector field.

Solution. Since $\phi' = \phi$ at any point, by the chain rule and (6.10),

$$\frac{\partial \phi'}{\partial x_j'} = \frac{\partial x_i}{\partial x_j'}\frac{\partial \phi}{\partial x_i} = a_{ji}\frac{\partial \phi}{\partial x_i},$$

whence the result follows. □

It is convenient to denote partial differentiation by use of a comma, thus:

$$\phi_{,i} \equiv \partial\phi/\partial x_i, \quad A_{i,j} \equiv \partial A_i/\partial x_j, \quad \text{etc.}$$

Problem 6.11 Show that the divergence of a vector field A_i is a scalar field.

Solution. The divergence of A_i is the quantity $A_{i,i}$. Transforming the cartesian coordinate system, we get for the divergence in the new system:

$$\begin{aligned} A'_{i,i} &\equiv \frac{\partial}{\partial x_j'}(A_j') = a_{ji}\frac{\partial}{\partial x_i}(A_j') = a_{ji}(a_{jk}A_k)_{,i} \\ &= a_{ji}a_{jk}A_{k,i} = \delta_{ik}A_{k,i} \\ &= A_{k,k}, \end{aligned}$$

which proves the result. □

Problem 6.12 Verify the vector identity (§ 3.3, iv):

$$\nabla.(\mathbf{a}\wedge\mathbf{b}) = \mathbf{b}.(\nabla\wedge\mathbf{a}) - \mathbf{a}.(\nabla\wedge\mathbf{b}). \tag{6.19}$$

Solution. Introducing the suffix notation,

$$(\mathbf{a}\wedge\mathbf{b})_i = \varepsilon_{ijk}a_j b_k,$$

and therefore the left-hand side of (6.19) is

$$(\varepsilon_{ijk}a_j b_k)_{,i} = \varepsilon_{ijk}(a_{j,i}b_k + a_j b_{k,i}). \tag{6.20}$$

Now,

$$(\nabla\wedge\mathbf{a})_i = \varepsilon_{ijk}(\partial a_k/\partial x_j) = \varepsilon_{ijk}a_{k,j},$$

and so the right-hand side of (6.20) is

$$b_i\varepsilon_{ijk}a_{k,j} - a_i\varepsilon_{ijk}b_{k,j} = \varepsilon_{kij}b_k a_{j,i} - \varepsilon_{jik}a_j b_{k,i},$$

where we have relabelled indices by the substitutions $i \to k, j \to i, k \to j$

in the first term, and by the interchange of i and j in the second term. Permuting suffixes in the ε symbols immediately brings this expression to the form (6.20), whence the identity (6.19) is verified. □

Problem 6.13 Prove the identity (§ 3.3, ix):

$$\nabla \wedge (\mathbf{a} \wedge \mathbf{b}) = (\nabla . \mathbf{b})\mathbf{a} - (\nabla . \mathbf{a})\mathbf{b} + (\mathbf{b} . \nabla)\mathbf{a} - (\mathbf{a} . \nabla)\mathbf{b}. \tag{6.21}$$

Solution. The left-hand side of (6.21) is

$$\begin{aligned}
\varepsilon_{ijk}(\mathbf{a} \wedge \mathbf{b})_{k,j} &= \varepsilon_{ijk}(\varepsilon_{klm}\, a_l\, b_m)_{,j} \\
&= \varepsilon_{kij}\,\varepsilon_{klm}(a_{l,j}\, b_m + a_l\, b_{m,j}) \\
&= (\delta_{il}\,\delta_{jm} - \delta_{im}\,\delta_{jl})(a_{l,j}\, b_m + a_l\, b_{m,j}) \\
&= a_{i,j}\, b_j - a_{j,j}\, b_i + a_i\, b_{j,j} - a_j\, b_{i,j} \\
&= \text{right-hand side,}
\end{aligned}$$

where use has been made of Problem 6.9(ii).

EXERCISES

1. Find the transformation matrix (a_{ij}) corresponding to a rotation of rectangular cartesian axes through 90° about an axis in the direction of $\mathbf{i}_1 - 2\mathbf{i}_2 + 2\mathbf{i}_3$.

2. Simplify the following expressions: (i) $\delta_{ab}\,\delta_{cd}\,u_{bd} - \delta_{ae}\,u_{ec}$, (ii) $\delta_{ab}\,\delta_{bc}\,\delta_{cd}\,v_d - v_a$, (iii) $\frac{1}{2}\varepsilon_{abc}\,\varepsilon_{abd}\,v_d$.

3. Using the suffix notation, verify the vector identity (§ 3.3, viii):

$$\nabla(\mathbf{a} . \mathbf{b}) = (\mathbf{a} . \nabla)\mathbf{b} + (\mathbf{b} . \nabla)\mathbf{a} + \mathbf{a} \wedge (\nabla \wedge \mathbf{b}) + \mathbf{b} \wedge (\nabla \wedge \mathbf{a}),$$

4. (*Quotient Law*) Let a_{ijk} be 3^3 numbers associated with each cartesian system, and let b_j be a vector (cartesian tensor of rank 1). Show that if $c_{ij} \equiv a_{ijk}\, b_k$ is a cartesian tensor of rank 2 for arbitrary b_j, then a_{ijk} is a cartesian tensor of rank 3.

(This result generalizes. If the contraction of a set of 3^n numbers $a_{i_1 \cdots i_n}$ with an arbitrary tensor of given rank $m (m \leqslant n)$ produces a cartesian tensor, then the a's form the components of a cartesian tensor of rank n.)

5. Show that the curl of a vector field A_i is a vector under proper orthogonal transformations.

6. A second rank tensor u_{ij} is *symmetric* if $u_{ij} = u_{ji}$, and *skew-symmetric* if $u_{ij} = -u_{ji}$. Show how an arbitrary second rank tensor a_{ij} may be expressed (uniquely) in the form $a_{ij} = s_{ij} + t_{ij}$, where s_{ij} is symmetric and t_{ij} is skew-symmetric.

Answers to Exercises

Chapter 1

1. (i) $xyz > 0$, i.e. the interiors of the four octants in which just one of the coordinates x, y, z is positive or all three are positive.
   (ii) $-1 \leqslant x-y \leqslant 1$, ($x, y$ not both zero). This is a slab with plane faces $x-y = \pm 1$, with the origin excluded.
2. (i) $4(3\mathbf{i}+\mathbf{j}+4\mathbf{k})$, (ii) $r^{-2}\hat{\mathbf{r}}$.
3. $-10/3$.
5. $2C\pi\sqrt{5}$.
7. (i) $x+y+z = a$, $x^3+y^3+z^3 = b$, (ii) $z(x-y)^2 = a$, $(x+y-2z)^2 = bz$.

Chapter 2

1. (i) $\frac{19}{21}$, (ii) $\frac{1}{21}(19\mathbf{i}+66\mathbf{j}+19\mathbf{k})$.
2. $\pi^3\sqrt{2}/24$.
3. $427/20$.
4. -4.
5. $\frac{2}{3}\pi(5\sqrt{5}-1)$.
6. $\frac{1}{6}\pi(5\mathbf{i}-3\mathbf{j})$.
7. $5/6$.
8. $3\pi/16$.

Chapter 3

1. $e^y - x\cos xy + 1$.
2. $x(x-2z)\mathbf{i}+y(y-2x)\mathbf{j}+z(z-2y)\mathbf{k}$.
3. (i) $2r^{-1}\sin r+\cos r$, (ii) 0.
4. (i) $2yz(3x^2+z^2)$, (ii) $2(\mathbf{i}+\mathbf{j}+\mathbf{k})$.
7. (i) 0, (ii) $2(x^2+y^2+z^2)^{-1}$.
8. $\phi = A\ln c+B$.

Chapter 4

1. πab.
5. πa^3.

Chapter 5

1. $-2\cos\psi$.

2. $r(\cos 2\theta \operatorname{cosec} \theta \, \hat{\mathbf{r}} - 3\cos\theta \, \hat{\boldsymbol{\theta}} + 3\sin\theta \, \hat{\boldsymbol{\psi}})$.

3. (i) $h_1 = h_2 = a\alpha, \quad h_3 = a \sinh u \sin v, \quad$ where $\alpha = (\sinh^2 u + \sin^2 v)^{\frac{1}{2}}$,
(ii) $a^2\alpha \sinh u \sin v \, dv \, d\psi, \quad a^2\alpha \sinh u \sin v \, d\psi \, du, \quad a^2\alpha^2 \, du \, dv$.

4. $\dfrac{\partial^2\phi}{\partial u^2} + \dfrac{\partial^2\phi}{\partial v^2} + c^2(\sinh^2 u + \sin^2 v)\dfrac{\partial^2\phi}{\partial z^2} = 0; \qquad \phi = A(u+v) + B$, where
A and B are constants.

5. $\alpha^{-1}[v(\cos\psi + u^2 \sin\psi)\mathbf{e}_1 + u(\cos\psi - v^2\sin\psi)\mathbf{e}_2] - \sin\psi \, \mathbf{e}_3$, where
$\alpha = (u^2 + v^2)^{\frac{1}{2}}$.

6. (i) $c^2(\sinh^2 u + \sin^2 v) \, du \, dv \, dz$, (ii) $uv(u^2 + v^2) \, du \, dv \, d\psi$.

Chapter 6

1. $\dfrac{1}{9}\begin{pmatrix} 1 & 4 & 8 \\ -8 & 4 & -1 \\ -4 & -7 & 4 \end{pmatrix}$.

2. (i) 0, (ii) 0, (iii) v_c.

6. $s_{ij} = \frac{1}{2}(a_{ij} + a_{ji}), \quad t_{ij} = \frac{1}{2}(a_{ij} - a_{ji})$.

Index